PREFACE

In the <u>Marietta Daily Journal</u> on March 1, 1987, Maggie Willis, in her weekly column, wrote: "Dr. Peter Inglis has a bee in his bonnet — a B-29 that is. He has gotten together a small task force to try to bring home one of the WW II bombers that were built at Bell Aircraft Corporation in Marietta. He has in mind a memorial, he said, 'to the designers, engineers, builders, flyers and victims'."

Writing this book in 1999/2000 is part of the process of getting that "bee" out of my bonnet. This book is not intended to be a documentary, nor a scrapbook, nor to glorify war. It is intended to keep the story and the memory of *Sweet Eloise* alive and to add to the B-29 chapter of Marietta's history book. It is intended to be a guide for history teachers, students, aviation and history buffs, local citizens, and Georgians across the state. It is a tribute to the millions of people around the world involved with B-29s during the war.

The setting of this unusual story is Marietta, Georgia, more specifically Dobbins Air Reserve Base (ARB). The story reaches into other states, communities, and into two combat areas. It is the story of the aircraft B-29 BW, *44-70113,* in combat, rescued, wrecked, restored, renamed *Sweet Eloise* and placed on display at Dobbins ARB. It includes the stories of the men who flew it and the efforts of Georgia Technical Schools and local community in bringing the ship from a state of disrepair in Florence, S.C. to its present condition and location. It includes the story of the Marietta Bell Aircraft Plant Number 6, the need for the B-29 in World War II, B-29 historical background, some aircraft data, statistics, specifications, and much more.

My interest in B29s began in 1943 with a family connection to the Bell Aircraft Plant in Marietta. It grew during my stint as a Flight Surgeon at Yokota Air Force Base (AFB) in Japan, 1947-49. Many will differ, but in my opinion the B-29 is the most important aircraft in our country's history. It was vital in World War II and negated a costly amphibious invasion of Japan. It was technically a difficult aircraft, the most expensive and deadliest weapon in our country's aircraft history, and part of the weapon system that gave the U.S. world dominance from 1944 though 1957. We became a Superpower. The small town of Marietta, Georgia played an important role in this impressive story.

The general populace, particularly the younger generation, is uninformed about the B-29. Thousands of people drive past *Sweet Eloise* every day and wonder, "What is that big old airplane over there?" "Why is it there?" and "How did it get there?" This book answers those three questions, and many more.

I have written this book to tell the story of Marietta's B-29, to relate some of her secrets, to tell about war-time Marietta, and to give an overview of the B-29 in general. She was a unique and interesting airplane. When you know her unusual history, you will know how and why she came to be at Dobbins. I hope you will want to visit her, take your children and grandchildren to visit her too, using this opportunity to give them a free history lesson, day or night. Go two miles south from the Big Chicken on Cobb Parkway, take a right at Dobbins ARB, a sharp right before the Security Gate into the parking area, and there you are. Take a walk into history and take a proud look at Marietta's impressive B-29, *Sweet Eloise.*

Pete Inglis

APPRECIATION

This book could never have been written, nor the displays displayed, without support and help from the following interested people. I can't name them all, but I would like to give special thanks to: Joe Bischof, Charlie Bollech, Louie Boos, Beau Clark, Chuck Clay, Harlan Crimm, Amy Whitney Dirnberger, Howard Ector, Bill Franklin, Bo Glover, Wanless Goodson, Francis Gore, Harry Ingram, "Possum" Hansell, Gilbert Johnson, Joe Kirby, Dempsey and Corene Kirk, Bob Mabry, Sherm Martin, Dot Mauldin, Jack Millar, Dick Morawetz, Bill Price, Pete Reeve, Judy Renfroe, Jack and Barbara Renshaw, C.G. Robinson, Dick Rogers, Bob Shearer, Tom Scott, Phil Secrist, Tony Serkadakis, Amadou Sey, Harold Shamblin, Coy Short, Regina Goldsworthy Stott, Jack Tetrick, Dick Wing, Steve and Virginia Tumlin, Sally Ann Wagoner, Maggie Willis, and Dr. Henry Zimmerman.

In addition I would like to give thanks to a few people who have been crucial to me: Jack Boone for his advice in publishing; Dan Cox for help on the museums and community support; Frank Duncan for layout, design, printing and binding; Mrs. Jane Alexander Eisele for the "Miss Jane's Country Diner" and "883rd Mess Hall;" Al Evans for being Mr. B-29 and for his book; Monroe King for help with the displays; Bill Kinney for thirteen years of help and support; Lt. Col. George Larson for two articles about our B-29, for his Forward, and for help with this book; Gov. Roy E. Barnes for his endorsement; Dr. Koji Yoda for exceptional translations of the leaflets and his continuing support.

My three children, Steve, Joye, and Jennie have been supportive and helpful. As a librarian, Jennie edited and created the Index. Steve and Joye have searched for typos. Especial thanks and credit goes to my wife, Ruth. We would not be living in Marietta if it had not been for her and the B-29. She has helped me with her writing and computer skills. We have surmounted many computer and other problems brought on by this book. She has been fine support, considering the fact that she frequently reminds me that "We did not retire for B-29s." Thank you, Ruth!

I could not have written this book, or been involved with the B-29, without you folks. It has been an interesting and memorable chapter in my life. Thank you, one and all, named and unnamed. I have enjoyed my association with each of you.

As the historian of the B-29 Superfortress Association, and while writing this book, I have had the opportunity of making contact with some heroes, and people who made a difference. Bob Bailey, Jack Boone, Joe Bischof, Beau Clark, Ray Clinkscales, Al Evans, John Frey, Harold Goldsworthy, Wanless Goodson, Francis Gore, "Possum" Hansell, Bob Hays, Gilbert Johnson, Bo Kincaid, Dempsey Kirk, Chester Marshall, Jack Millar, Dick O'Hara, Bill Price, Pete Reeve, Jim Reifenschneider, Bill Rohmer, Joel Rutledge, Louis Sohn, Tony Serkadakis, Rudy Stankus, Jack Tetrick, Ed Wigley, Richard Wing, and Henry Zimmerman. Each of you men did your "bit" a half a century ago. Thank you. Tom Brokaw would be proud of each of you.

Pete Inglis

FORWARD

My involvement with Pete Inglis and his efforts to tell the story of a single Boeing B-29 Superfortress began during World War II. My father, George W. Larson, was assigned to the 135th United States Naval Construction Battalion (Sea Bees), which on Tinian Island helped build the four 8,500 foot long runways on North Field. This airfield was used by B-29s to bomb Japan, including *Sweet Eloise*. Pete Inglis contacted me because of my writings about Tinian Island, B-29s and World War II subjects. I wrote two articles about this B-29, telling its story of resurrection from neglect to display. The story continues today as more plans are in the future for this Marietta, Georgia based aircraft. Marietta was the site of a large B-29 production facility during World War II, run by Bell Aircraft Corporation. The story of *Sweet Eloise* is history, one which does not get told as often as it should be. The Bell Aircraft B-29 production lines tie the war years of 1941-1945 to Lockheed and the present. Pete Inglis tells a story of the B-29's service in World War II, its discarding after the war, saving from the scrap yard, restoration, and display. This B-29 has been saved for future generations, but there is much more to do. There is not a single museum in the United States which is solely devoted to the B-29. Everyone can learn about this Superfortress from this book and perhaps contribute to the building of a B-29 museum. As a World War II historian and writer, the B-29 was part of my family history, and now, we hope it will become one of your interests. The book tells the personal side of combat and the importance of a rare warbird, *"Sweet Eloise."*

Lt. Col. George A. Larson, USAF (Ret.)
World War II Historian and Writer
Rapid City, South Dakota

TABLE OF CONTENTS

SECTION I

Chapter 1 ———————
Sweet Eloise in Combat ——————

- Origin
- Stepping Stones
- Gen. Haywood "Possum" Hansell
- Capt. Ray Clinkscales
- Lt. Norman Adamson

Origin

To begin the story of Marietta's B-29, we start with her production at Wichita, Kansas. *Sweet Eloise,* was not made at the Bell Aircraft plant in Marietta. She was designated as 44-B-29-A-BW, (a 19<u>44</u> contract, a <u>B-29A</u> model, distinguished from a stripped-down B model, and built at <u>B</u>oeing– <u>W</u>ichita). A B-29 Superfortress is the "big sister" of the B-17 Flying Fortress, and "little sister" of the B-47 and B-52, all Boeing aircraft. In May 1945, after testing, she was delivered to Kearney Army Airfield, Nebraska, dedicated, but not named.

Alton Evans and Col. William Kane after installing Data Plate under left elevator 1998

Her official data plate, or "ID," is located under the left elevator. This was re-installed on October 6, 1998, by Col. William Kane, Commanding Officer, Dobbins ARB, and by Mr. Alton Evans. Al Evans had removed the data plate at Florence, S.C. in 1973 for safekeeping prior to aircraft restoration.

Sweet Eloise not only flew more missions (27) in combat than the average B-29, but she was a "Sweet Airplane," always bringing her crew home safely, and without injury. A few B-29s flew more than 60 missions. Her two principal crews also flew more missions than most, in different B-29s. She always accomplished her mission, she was a fast airplane, with no "aborts," (flights aborted for mechanical problems or damage), with no emergency landings on Iwo Jima. She was a "lead" airplane. She and her crews performed like thoroughbreds.

Stepping Stones

In May 1945, the new B-29 was dedicated at Kearney, Nebraska, assigned to Lt. Gordon Burgess. With a crew of eleven she was flown to San Francisco, Pearl Harbor, Johnston Atoll, Kwajalein Atoll, and finally to Saipan, an island in the Mariana Islands, a distance of about 7,000 miles from home. The Mariana Islands (Guam, Tinian, and Saipan) had been conquered by the Army, Navy, Marines, and Army Air Corps. There was still heavy fighting on Iwo Jima and Okinawa.

Saipan was part of the costly "stepping stone" or "island hopping" strategy Admiral Chester Nimitz planned, in preparation for the invasion of Japan. The Mariana Islands are 1,600 miles from targets in Japan, within striking range of the B-29s. The island of Saipan is an idyllic island paradise with a near-perfect climate, about four by twelve miles in size.

Two divisions of Marines and 16,000 Army troops captured Saipan from the heavily fortified Japanese after vicious fighting (from June 11 to July 8, 1944). The greatest suicidal banzai attack in recorded history, which consisted of 5,000 fanatic Japanese troops, was defeated by U.S. troops. The cost was heavy, with over 3,500 Americans killed

and 15,000 wounded. An estimated 25,000 - 30,000 Japanese were killed.

Four million cubic yards of coral and rock were moved by the Sea Bees and Aviation Engineers to hastily build Isley Field on Saipan. Isley Field consisted of two parallel 8,500 foot runways, 200 feet wide, with 125 hard-stands. At the end of the runway there was a drop to the ocean which, like an aircraft carrier, permitted the straining bombers to take-off with heavier loads of bombs and fuel.

Tinian Island, three miles across the Saipan channel, was captured in nine days, and became the busiest airport in the world at that time, including operations by *Enola Gay* and *Bocks Car.* Tinian had four parallel runways to accommodate hordes of B-29s. This allowed two B-29s to take off every forty-five seconds. The fighting for Guam, 125 miles to the south, lasted for 19 days, until August 10, 1944.

Fleets of these "Dreadnaughts of the Air" began to mount the terrible, intense bombing of Japan from these three bastions for eight months until unconditional surrender was finally attained on August 15, 1945. This helped prove that General Billy Mitchell was vindicated from his court-marshal in 1924, for predicting air power could defeat an enemy without an invasion. At that time, General Mitchell also predicted that on some quiet Sunday morning Japan would bomb Pearl Harbor. Our bombing campaign wrought on Japan exceeded the devastation wrought on Hitler's Europe by our air power.

Tinian Island is where the B-29s *Enola Gay* with Col. Paul Tibbets in command, and *Bocks Car* with Maj. Chuck Sweeney, took off to drop their atomic bombs on Hiroshima and Nagasaki. The name "Enola Gay" is Col. Tibbet's mother's maiden name. The name "Bocks Car" was derived from Maj. Fred Bock, her first commander. There is a small B-29 museum on Tinian Island in memory of these momentous events. There is no other museum in the world dedicated solely to B-29s.

The three Mariana Islands were poised like daggers pointed at the heart of Japan. With these three huge bases on the Marianas and the B-29 Superfortress, experts predicted it was just a question of time and diligence before Japan surrendered.

Upon her arrival on Saipan in May 1945, our super-bomber was assigned the number *Z Square 58* indicating she was the 58th plane in the 500th Bomb Group, all of which bear the letter "Z" on their tails. She was assigned to the 883rd Bomb Squadron. The "Square" indicates the 73rd Bomb Wing (the Square was later discontinued). Saipan was exclusively 73rd Bomb Wing territory.

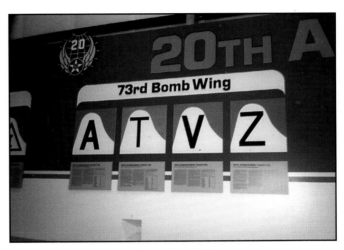

Group tail markings of the 73rd Wing:
A - 487th Group
T - 488th Group
V - 499th Group
Z - 500th Group-
500th Group has 3 Squadrons: 881st; 882nd; 883rd
Each Squadron has 20 aircraft
Sweet Eloise, Z-58, is 58th in the Group
   Pima Air Museum, Tucson

The commanding officer was Lt. Col. William McDowell who lives in Brady, Texas. The Flight Surgeon for the 883rd was Capt. Henry Zimmerman who now lives in Vero Beach, Florida. "Doc Z" is an eminent retired cardiologist. He earned a Purple Heart while flying on one of his six missions over Japan for which he had volunteered as an observer. He served on Saipan for 15 months. He wrote the first medical article on B-29 combat fatigue, based on his original research with the 883rd Squadron. This was his first of about 120 medical articles. "Doc Z" attended our dedication here in 1997 where he saw some of his old patients (former crewmen).

The 73rd Bomb Wing was commanded by Gen. Emmett (Rosie) O'Donnell. The 21st Bomber Command was commanded by Gen. Curtis LeMay whose predecessor was Gen. Haywood "Possum" Hansell, Jr. The 20th Air Force was commanded by

General Henry F. "Hap"Arnold. Gen. George C. Marshall commanded the Joint Chiefs of Staff. The Commander-in-Chief was President Harry S. Truman, who succeeded President Franklin D. Roosevelt at his death on April 14, 1945.

Maj. Gen. Haywood "Possum" Hansell, Jr.

Maj. Gen. Haywood "Possum" Hansell, Jr. fits into this story because he felt strongly that Marietta needed a B-29. He had roots in Marietta. He lived in Atlanta and Roswell, visiting Marietta frequently. He was the brother-in-law of Mariettan, Mrs. Brooks (Lucia) Smith, and the uncle of lawyer, Hansell (Hap) Smith. "Poss" died in 1993. He graduated from Georgia Tech. His nickname came from classmates at University of the South who thought he resembled an opossum. His grandfather, Gen. Andrew Hansell, was Adjutant General for Georgia, and he built the mansion Tranquilla on Kennesaw Avenue in Marietta and Mimosa Hall in Roswell. They were a prominent family in Marietta, and there is a Hansell Street in downtown Marietta.

"Possum" was the Commanding Officer of the 21st Bomber Command, prior to Gen. Curtis LeMay. "Possum" wrote two fine books, The Air War Plan That Defeated Hitler and Strategic Air War Against Japan. These two books are still used by the Air War College. He practiced strategic bombing, picking the targets that were most vital to the war effort. He thought Marietta should have a B-29 on display, since they were built here, and that we should have a 20th Air Force Museum here, since the 20th AF was born here. His friend, "Curt" LeMay concurred. Savannah has the impressive fourteen million-dollar 8th Air Force Museum,

since the "Mighty 8th" was born in Savannah.

In January '45 Gen. Hansell was replaced by General LeMay because "Possum" persisted in performing high altitude pin-point "civilized" bombing for which the aircraft was designed, with unsatisfactory results. He emphasized strategic bombing of vital targets such as aircraft plants, refineries, electrical production, steel plants, shipyards, harbors, transportation, communication, etc. When Gen. LeMay took command, he initially continued the mass, high altitude, daytime bombing in the conventional manner.

This was not as destructive as the incendiary napalm low altitude carpet-bombing of Japanese cities at night, which he instituted. Gen. LeMay felt that this low altitude carpet-bombing was justified by the number of cottage industries in Japan, in which nearly every home was producing a piece for the war effort, as well as to demoralize the populace. LeMay's strategy paid off faster. The great outpouring of this nation's treasure, manpower, and material were now coordinated to strike the final blow against Japan in retaliation for Pearl Harbor and to halt her imperialistic conquests.

8th Air Force Museum in Pooler, Ga. near Savannah

Bugs Bunny certificate Army Air Corps, from Richard Wing

On Saipan, in May of 1945, *Z-58* was finally ready for combat. Training a B-29 pilot required 27 weeks, a navigator 15 weeks, and a gunner 12 weeks. The melding of equipment, training, and organization made this superb aircraft ready to perform its mission. The entire crew of 11 men was brought together for the first time, then more time was required for integration of the crew.

Each man had been through intensive training in each specialty, at various bases primarily in the Mid-West, (especially Kansas and Nebraska) such as: Great Bend, Smokey Hill, Salina, Topeka, Wichita, Walker, Pratt, Herrington, Hays, and Russell in Kansas; Lincoln, Harvard, Kearney, McCook, Fairmont, Grand Rapids, Grand Island, and Gowen in Nebraska; Warner-Robins in Georgia; Ft. Meyers in Florida; Pyote and Dalhart in Texas; Denver, and Colorado Springs in Colorado; Fresno in California; Clovis in New Mexico; Tucson and Kingman in Arizona; and many others, and sometimes involving simulated bombing runs on Cuba, Jamaica, or Puerto Rico.

There were many forced landings and B-29 crashes throughout the southern tier of states during this period. The aircraft also underwent a myriad of major modifications to become combat-ready. The entire program was in chaos for much of 1943-44. This program was jokingly called the "Battle of Kansas" by some of the new B-29 personnel. Prior to the B-29s being flown to India and China, extensive modifications needed to be made to each aircraft.

Capt. Ray Clinkscales

On Saipan Capt. Ray Clinkscales and his veteran crew of 11 men were transferred to *Z-58, Marilyn Gay,* now renamed *Sweet Eloise,* and they flew her for eight missions over Japan. Capt. "Clink" was a veteran pilot having commanded *Z-53 Ancient Mariner* on 22 other distinguished missions, which included shooting down five fighters, two forced landings on Iwo Jima for battle damage to the bomb bay doors in a dramatic "buddy battle," (page 13), and for fuel. He and his crew accumulated 464 combat hours involving his 30 missions. It is unknown how many hours he had during his first tour in Europe and Africa, flying a B-24 Liberator and A-20 Havoc. This was his second tour.

Clinkscale's targets in *Sweet Eloise* were:

May 23 1945 Tokyo	June 5 Kobe
May 26 Tokyo	June 7 Osaka
May 29 Yokohama *	June 10 Hitachi
	(Tokyo suburb)
June 1 Osaka	June 14 Osaka

* Gunner Richard Wing relates that on this raid two incendiary napalm bombs hung up in the bomb bay. Two crewmen climbed into the bomb bay and released them over the ocean, when it was safe to open the bomb bay doors. (see page 13)

DATE	MISSION NUMBER	TARGET	AIRPLANE NUMBER	RESULT	SORTIES INDIV	SORTIES TOTAL	COMBAT HRS. INDIV	COMBAT HRS. TOTAL	INITIALS	E/A CLAIMS KILL	E/A CLAIMS PROB	E/A CLAIMS DAM	REMARKS
4-22-45	FE-3	Okinawa	42-65256	-	1	18	12:55	290:00	RCVER	0	0	0	Fighter Escort
4-24-45	53	------	44-69868	-	0	18	1:55	291:55	RCVEB	0	0	0	Mech Failure
4-26-45	56	Kyushu	42-65296	P	1	19	14:05	306:00	RCVEB	0	0	0	
5-5-45	65	Kure	42-65296	P	1	20	16:20	322:20	RCVEB	0	0	0	
5-14-45	69	Nagoya	44-69878	P	1	21	15:05	337:25	RCVEB	0	0	0	
5-18-45	FE	------	42-94001	-	1	22	12:15	349:40	RCVEB	0	0	0	Fighter Escort
5-23-45	72	Tokyo	44-70113	P	1	23	14:15	363:55	RCVEB	0	0	0	
5-25-45	73	Tokyo	44-70113	P	1	24	14:50	378:45	RCVEB	0	0	0	
5-29-45	74	Yokohama	44-70113	P	1	25	13:55	392:40	RCVEB	0	0	0	
6-1-45	75	Osaka	44-70113	P	1	26	15:00	407:40	RCVEB	0	0	0	
6-5-45	76	Kobe	44-70113	P	1	27	14:35	422:15	RCVEB	0	0	0	
6-7-45	77	Osaka	44-70113	P	1	28	14:30	436:45	RCVEB	0	0	0	
6-10-45	78	Tokyo	44-70113	P	1	29	14:15	451:00	RCVEB	0	0	0	
6-14-45	79	Osaka	44-70113	P	1	30	13:50	464:50	RCVEB	0	0	0	

**Capt. Ray Clinkscales' Partial Combat Record, Note: Final 8 Missions were in *Sweet Eloise*
From S/Sgt Richard Wing**

JUNE 5, 1945

B-29s FIRE KOBE, GREAT JAP PORT

450 Battle Snow and Fog, Fighters, but Hit Targets

Guam—*AP*—Approximately 450 Superfortresses battled through snow, fog, thunderheads, accurate anti-aircraft fire and fairly strong fighter opposition today to transform the industrial and transportation center of Kobe into a mass of smoke and flame.

The B-29s, flying through weather so bad they had to make the trip without fighter escort, found perfect weather over Japan's largest port city and spent an hour setting it afire with 3,000 tons of incendiary bombs in a blazing first anniversary celebration of the initial Superfort raid.

Returning pilots described the results as excellent. They guessed damage would exceed the nearly seven square miles burned out of Yokohama in the May 29 strike at Nippon's second largest port.

About 40 Japanese fighters, some of them fairly aggressive, were sent up to intercept the tight bomber formations, and a few flew above the B-29s to drop phosphorous bombs. But they could not keep the daylight attackers from coming in dead on their target in the third fire strike at Japan's sixth largest city.

Six hours after the last bomber turned away Japanese broadcasts conceded that fires were only "gradually being extinguished." Tokyo said fire bombs also set blazes in Mikageachmi and Ashiya, industrial towns respectively two and six miles east of Kobe on the rail lines running along Osaka Bay to the city of Osaka, last previous B-29 target.

The Domei news agency, quoting a joint army-navy communique issued in Kobe, claimed 56

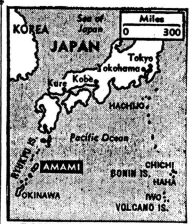

BOMB TARGET—Map shows location of Kobe, Japan, heavily damaged by B-29s raid.

superforts were shot down, and 144 "heavily damaged." Seven Japanese interceptors were listed as lost, "including those that carried out body-crashing attacks."

Superforts centered their incendiaries on a heavily industrialized square mile area of eastern Kobe, including the city's two main railway stations, the extensive Kobe steel works, shipyards and major docks. This is approximately adjacent to nearly nine square miles in the center of the

See JAPAN Page 3

city burned out in previous strikes.

While the B-29s were enroute back from Kobe, Maj. Gen. Curtis E. Lemay's 21st bomber command headquarters reported that nearly three and one-half square miles of industrialized Osaka was destroyed by a 450-plane incendiary raid Friday. This increased to 11½ square miles the extent of B-29 wrought destruction in that city.

Reconnaissance photographs showed 11 industrial targets heavily damaged. Those targets included harbor installations, drug and dyestuff manufacturing plants, a metal industry factory, where aircraft propellers were produced; iron works and other manufacturing plants.

Superforts centered their incendiaries on a heavily industrialized square mile area of eastern Kobe, including the city's two main railway stations, the extensive Kobe steel works, shipyards and major docks. This is approximately adjacent to nearly nine square miles in the center of the city burned out in previous strikes.

The attack was concentrated in an industrial section along Osaka harbor south of the Yodo River.

The B-29s attacked Kobe in daylight, as they did on their first incendiary raid on Kobe, Feb. 4. The second fire raid, March 17, which burned out shipbuilding yards was made in early morning darkness. Kobe also was given a 150-plane demolition raid May 11.

The Imperial Government railway shops, the main Sannomiya railroad station and the East Nada railroad yards were among the communications centers attacked today.

Big factories, concentrating on shipbuilding, marine engines, steel, electrical equipment and machinery, are crowded into industrial areas against which the B-29s loosed bombs.

Kobe is about 250 miles south-

New York Times. The Associated Press from F/O Francis Gore

PARIS EDITION

THE STARS AND STRIPES

Daily Newspaper of U.S. Armed Forces   in the European Theater of Operations

The Weather Today
PARIS & VICINITY
Scattered clouds, max. temp.: 82
STRAITS OF DOVER
Partly cloudy, max. temp.: 76

The Weather Today
RIVIERA
Clear, max. temp.: 83
GERMANY
Scattered clouds, max. temp.: 78

Vol. I—No. 318   1 Fr.   1 Fr.   Sunday, June 10, 1945

Ninth Army Nears End Of ETO Job

By Ernie Leiser
Stars and Stripes Staff Writer

WIESBADEN, June 9.—Lt. Gen. William H. Simpson's U.S. Ninth Army is slated to close up its ETO operations about June 15, 12th Army Group Headquarters revealed today.

Army Group officials did not disclose what the Ninth's new assignment would be. Presumably, however, it will follow Gen. Courtney Hodges' U.S. First Army to the Pacific as America shifts the full weight of its armed might against Japan.

Whether Simpson will continue as chief of the army he led across Germany to the Elbe has not yet been announced.

The Ninth, which less than a month ago was the world's biggest army with 21 divisions, five corps and approximately a million men, has been shifting its units rapidly to the south during the last weeks, and turning a large number of them over to other armies.

Held Fourth of Germany

When Simpson's command took over the First Army area after Hodges had returned to the States on his way to the Pacific, the Ninth Army occupied nearly a fourth of Germany, but little of it was in the general area indicated for U.S. occupation.

Subsequently, 12th Army Group has ordered most of the Ninth's major units south. This shift, which has been achieved "much faster than expected," according to plans officers, has reduced the American area by more than 50,000 square miles.

Pending announcement of the Allied and Russian occupation zones, the British have been relieving the Ninth Army troops in most

(Continued on Page 8)

Hopkins Here From Moscow

Harry Hopkins, emissary of President Truman to Marshal Stalin, is in Paris with Mrs. Hopkins. U.S. circles said Hopkins would leave shortly for Washington.

Informed sources in Paris said Hopkins was returning from a successful mission in Moscow. He was reported to have Stalin's plans for the coming conference with President Truman and Prime Minister Churchill and outlines of methods to settle the Polish situation and for control of Germany and Austria.

The breaking of the deadlock over the veto formula at the San Francisco Conference was attributed to Hopkins' talks with Stalin.

En route to Paris, Hopkins stopped in Berlin, where he was a guest of Marshal Zhukov and where he visited the underground refuges beneath Hitler's chancellery.

Miracle on 44th St.

3 B29 Fleets Hammer Japan; Yanks Blast at Okinawa Pockets

Ground crews at a Marianas base service a fleet of Superfortresses just returned from a strike at a Japanese homeland target. The big silver bombers which have turned many Japanese cities into smoking ruins get a thorough going-over between missions.

600 ETO-ers, Rich in Points, Arrive in U.S.

By Ben Price
Stars and Stripes Staff Writer

CAMP KILMER, N.J., June 9.—Six hundred combat soldiers from the 12th Army Group, most of them with 95 points or more, arrived in the U.S. today on their way to separation centers and probably civilian life.

They make up the first ETO group to be sent home for discharge under the Army's point system.

Thus, three of the four phases of the point system have been completed for the lucky 600. The first began 24 days ago when these GIs were called from their bivouacs in Germany, had their service records, and other papers put into shape, and were given orders for the journey home.

The second phase was three days of processing at the staging area near Le Havre. The third phase was the boatride, the long sweating-out period in which the GIs "refought" the war, ate good food and began to think about how it would be when they had switched from ODs to tweeds.

The fourth and last leg will be the separation center, where each eligible soldier will get an honorable discharge.

Pvt. Lester Greenberg, of New York, 95-point ex-member of the 258th FA Bn., hopes there won't be any delay in getting out. In fact, he has been worried about it all the way from Europe.

"This is too good to be true," said Greenberg. "They advertise these separation centers as being able to discharge you within 48

(Continued on Page 8)

Full Surrender Demand Bars Jap Peace Plea, Premier Hints

NEW YORK, June 9 (AP).—A special session of the Japanese Imperial Diet, ordered by Emperor Hirohito and opened today by him, was told by Premier Suzuki that the Allied demand for unconditional surrender leaves "no alternative but to fight to the very last."

Although he acknowledged the current war situation as "the gravest crisis in Japan's history," Suzuki said he was convinced that "the enemy will be smashed in decisive battles on our homeland."

He said that "no one in the world more earnestly desires world peace and welfare of humanity than Emperor Hirohito," and he asserted that "in the final analysis this war is one of liberation against an Anglo-American scheme for wholesale and permanent enslavement of the East Asiatic peoples."

Suzuki said that Japan was determined "to act in unison with her allies until ultimate victory." He said his country desired to promote "friendly relations with neutral powers.

"Japan's policy," he continued, was "to let every nation in the world enjoy its proper place in the sun, free from aggression, and to enhance justice for all humanity."

Trieste Pact Signed by Allies

LONDON, June 9 (AP).—The Foreign Office announced an agreement on the temporary military administration of the territory of Venezia - Giulia, which includes Trieste, has been reached by Great Britain, the U.S. and Jugoslavia.

(The United Press also reported the agreement had been announced by the State Department in Washington.)

Officials said that military details

Superforts' Attack On Plane Plants Is 1st Triple Raid

GUAM, June 9 (ANS).—Three fleets of Superfortresses today blasted Japanese aircraft plants at Osaka, Nagoya and Kobe as for the first time the 21st Bomber Command dispatched its Marianas-based B29s to attack more than two targets.

Each of the fleets numbered between 100 and 150 B29s, which carried high explosives only—the first mission in three weeks that dropped no incendiaries.

(The bombers dropped 750 to 1,000 tons in each strike, the United Press said, as B29 strategy returned to precision bombing of key industrial targets after blanket incendiary attacks.)

(The UP quoted the Japanese radio as saying that 5,000,000 or more Japs had been left homeless by B29 raids in May on five major cities. The broadcast, which did not refer to damage of B29 raids in June on Osaka and Kobe, admitted that in Tokyo alone more than 3,000,000 persons were homeless.)

Osaka, Nagoya and Kobe are Japan's second, third and fifth largest cities, respectively. They lie in a 100-mile northeast-to-southwest line west of Tokyo, on the main homeland island of Honshu. It was the third raid in three months on Osaka, the greatest arsenal city in the Orient; the fourth of the war on Kobe, and the 14th on Nagoya, Japan's aircraft center.

The planes were over their targets, which included five large aircraft plants in the three cities, during daylight. One of the objectives was the Kawanishi plant at Naruo, at Osaka's outskirts, where Japanese Navy fighters are made. More than a third of the plant was reported destroyed in a raid Jan. 19.

Twice before Superforts have hit two targets in a single raid but never three. On April 7 Tokyo and Nagoya were bombed, while on April 16 Tokyo and Kawasaki were hit.

There was no official report on results of today's attacks or of B29 losses. Earlier this week, 21st Bomber Command headquarters reported that 90 square miles of Japan's industrial centers had been destroyed or burned out.

Army to Pare Critical Score

WASHINGTON, June 9 (ANS).—The final critical point score governing the discharge of men from the Army will be slightly less than the 85-point interim score already announced, the War Department disclosed today.

Revealing that the critical score would be announced sometime next month, the Department said that the reduction in the required number of points for eligibility for discharge could not be "expected to be very great."

It is impossible to forecast what the final score will be, it was stated, since an analysis must be made of reports now being received from the various theaters of operation. (The ETO announced Thursday that it had completed its survey of adjusted service rating cards six days ahead of schedule.)

The 85-point score represents credits which accrued up to May 12. At some future date a new time will be fixed and soldiers who have reached the critical score since May 12 then will become eligible for separation, the Associated Press said.

Yanks Open Heavy Assault On Jap Pockets on Okinawa

GUAM, June 9 (ANS).—Supported by planes, warship bombardment and artillery, American soldiers and marines have launched a heavy assault on the two remaining Jap pockets on Okinawa, where the enemy occupies 19 of the island's 485 square miles. Onc

(Continued on Page 5)

13 Soldiers Punished for Abuse Of Vet Patients, Inquiry Told

B-29 Fleets Hammer Japan The Stars and Stripes from Paul Baker

Strike photo of Kure   by Army Air Corps   from Francis Gore

Ancient Mariner **dropping 500 lb bombs   by Army Air Corps**

Supine Sue drops 184 napalm bombs on Yokohoma, piloted by Capt. Clinkscales
Photo by gunner Sgt. Howard Clof, pilot Col. Pete Reeve
Sweet Eloise was in this formation but is not visible,
by Army Air Corps, from Col. Pete Reeve

Ruins of Tokyo
Photo by Wanless Goodson

Ruins of Yokohoma
Photo by Wanless Goodson

***Sweet Eloise* en route to Japan
by Army Air Corps**

***Sweet Eloise* lifting off from Saipan en route to Japan
from Al Evans Collection**

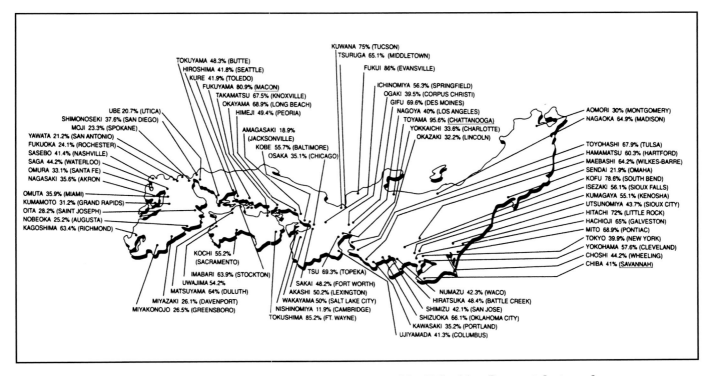

Bombed-out Japanese cities paired with comparable U.S. cities, Per-cent destroyed
Smithsonian Institution

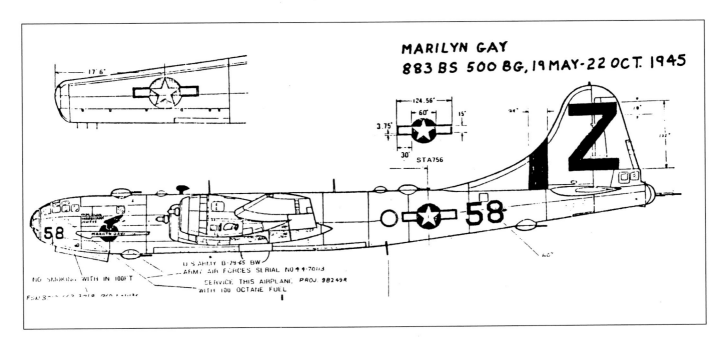

Markings of *Sweet Eloise (Marilyn Gay)* from S/Sgt. Richard Wing
Note: On the nose, Ball and Spear insignia of the 73rd Bomb Wing

S. J. GUNNER BRINGS DOWN 2 NIP PLANES

TOP GUNNER ON Z SQUARE 58 WITH 3 CONFIRMED KILLS. + 3 PROBABLE KILLS. FOR THIS MISSION, CAPT. CLINKSCALES & MYSELF RECEIVED THE DISTINGUISHED FLYING CROSS.

To become a member of the B-29 Superfortress Association, Inc., please complete the form below and mail to:

**B-29 Superfortress Inc.
P.O. Box 811
Atlanta, Georgia 30301**

Name RICHARD. C. WING

Address 3770 ERIN DR

City SPARKS State NV Zip Code 89436

702- ~~425~~ 425-4476

Day Phone   Evening Phone

702-425-4476

**Richard Wing joins
Marietta B-29 Superfortress Assn.**

TWENTIETH AIR FORCE, Pacific Headquarters.—Sgt. Richard C. Wing of San Jose, gunner on a Marianas-based B-29, was officially credited with shooting down two Jap fighter planes in a furious air battle over Tokyo.

The battle began after the B-29s had dropped their bombs on the important Mushashina aircraft plant. Flyers on the mission said that Jap anti-aircraft gunners "threw up everything they had" and that Jap fighter planes, ganging up to make dozens of attacks, followed the formation far out to sea.

BRINGS DOWN TWO

Sergeant Wing's B-29, undamaged in the long battle, dropped out of formation to protect another Superfortress which had been badly hit and forced to lag behind with a blazing engine. As Jap pilots swarmed in to try to knock the crippled bomber out of the air, Sergeant Wing shot down two of them while other gunners destroyed two more and probably destroyed a fifth.

About 60 miles out from Tokyo, the gunners ran out of ammunition just as the only remaining Jap fighter started to make a final attack. The pilot chased the fighter away by diving on him, and both B-29s returned safely to their base.

Sergeant Wing lives at 1187 Peach Court, San Jose.

Sergeant Wing is the son of James Wing, deputy U. S. marshal and former Santa Clara police officer, and Mrs. Hazel Wing of 1072 Washington St., Santa Clara.

**Article about Richard Wing
in <u>San Jose Mercury News</u>
from S/Sgt. Richard Wing
See page 13**

CAPT. RAY CLINKSCALES CREW
from S/Sgt. Jim Reifenschneider

Ray Clinkscales	Tom Benwell	Bob Spicer	Ed Struble	Jim Wride	Bob Kahn
Pilot	Co-pilot	Navigator	Bombardier	Flight Engnr	Radar
Belmont, N.C.		Lake Placid, FL		Bartlesville, OK	

Francis Sobeck	Richard Wing	Don Chambers	Jim Reifenschneider	Elmo Glockner
Radio	Gunner	Gunner	CFC Gunner	Tail Gunner
	Sparks, NV	New Philadelphia, OH		

LEAD AIRPLANE
30 MISSIONS OVER JAPAN,
8 IN *SWEET ELOISE*
DISTINGUISHED FLYING CROSS
3 JAPANESE FIGHTERS SHOT DOWN
6 AIR MEDALS

Author's notes: S/Sgt. Richard C. Wing, DFC CFC, sent me the below note and story from the Stars and Stripes. This event occurred on April 7th. Capt. Clinkscales was piloting *Z-53, Ancient Mariner,* in which Dick Wing was one of the gunners. The disabled B-29 was piloted by Lt. Norman Adamson, flying *Z-48, Million Dollar Baby,* not *Z-49, Three Feathers.* Adamson's crew may owe their lives to Clinkscales. Capt. Clinkscales and Lt. Adamson both became future pilots of *Sweet Eloise,* Clinkscales starting on May 23, and Adamson starting on June 1. The entire Clinkscales crew was awarded a Distinguished Flying Cross for this act of heroism and skill. See page 11.

Some memorabilia from Stars and Stripes, Marianas, 1945

"SO YOU WANTA RAM?"

Here's how to scare away Jap fighters when nothing else will. Capt. Ray Clinkscales was high-tailing his B-29 homeward after a Tokyo run. Some ways out, his Radio Operator, Francis Sobeck, picked up a distress signal from another B-29, which had dropped out of formation, crippled. To protect the damaged Superfortress, Clinkscales dropped behind to protect and escort him. He flew cover through dozens of Jap fighters. In the wild running battle that followed, Clinkscales' gunners shot down four Jap fighters and possibly destroyed a fifth.

But they kept on the B-29's tail, until finally 60 miles out from Tokyo, only one Jap fighter still persisted. At this point, the B-29 ran out of ammunition. The fighter must have known it because it did a few loops, sort of like a baseball pitcher winding up, and got set to come in for a kill. But he didn't know what Clinkscales had in mind for him. Maybe the B-29 pilot didn't know himself.

But suddenly he turned the SuperFortress in the direction of the attacker. The Jap then paused at that one. Then suddenly the B-29 lurched forward as though to ram the little plane. Surprised, the Jap turned tail and putting on high speed, headed for home. The two B-29s arrived back to the Marianas safely."

"As I recall, the other plane was piloted by the Captain of *Three Feathers* crew and landed at Iwo Jima, safe." Richard C. Wing

---

Upon completion of these eight missions, Capt. Clinkscales was returned to the states. He remained in the Air Force and retired as a Colonel. He presently resides in Myrtle Beach, S.C. He visited Marietta for the dedication of *Sweet Eloise* on May 6, 1997 where he was recognized, as was his gunner, S/Sgt Jim Reifenschneider. This crew received two Distinguished Flying Crosses and six Air Medals, and one Distinguished Unit Citation. *Marilyn Gay* was also the Lead Ship. One of her crewmen, Sgt. Richard Wing, shot down three Japanese fighters, and possibly three others, on one mission.

REUNION OF CLINKSCALES CREW 1995
Jim Wride   Jim Reifenschneider   Ray Clinkscales
Bob Spicer   Richard Wing
By Corene Kirk

Lt. Norman Adamson

Lt. Norman Adamson commanded the next crew from June 1 to August 28. This crew had flown 17 missions in other B-29s, then flew 13 missions in *Z-58* for a total of 30, then three POW (Prisoner Of War) mercy missions in which they dropped supplies on American POW camps after the Armistice. They were the lead crew in their squadron for much of this time, which authorized them to carry a stripe around their aft fuselage. This aircraft was named after Lt. Adamson's baby daughter, *Marilyn Gay* (no relationship to *Enola Gay*). She never had any nose art until the Cold War.

One of her crewmen, the tail gunner Sgt. Joe Bischof, kept a diary and letters to his mother, portions of which are printed later in this chapter. The co-pilot, Lt. Wanless Goodson, who now lives in Welch, West Virginia, visited Marietta for the ceremony of changing her name to *Sweet Eloise* in 1994, as did her radar operator, Cpl. Jesse Colvin, from Pulaski, Tennessee. Goodson pointed out a large aluminum patch about 10 X 10 inches

1ST LT. NORMAN ADAMSON CREW
from F/O Francis Gore

Top row

1st Lt. Norman Adamson	2nd Lt. Wanless Goodson	2nd Llt. John Hartnett	F/O Francis Gore
Pilot	Co-Pilot	Navigator	Bombardier
Seattle, WA	Welch, WV	New York, NY	N. Hornell, NY

Middle row

T/Sgt Ralph Barnett	Cpl. James Layden	Cpl. Wm. Baldwin	Pfc. Stanley Convissor
Flt. Engineer	Radio	Gunner	Gunner
Miami, FL		Charleston, WV	Brooklyn, NY

Bottom row

Cpl. Gerald Bartscherer	Cpl. Jesse Colvin	Cpl. Joe Bischof
Gunner	Radar	Tail Gunner
Denver, CO	Pulaski, TN	New Bergen, NJ

R E S T R I C T E D

HEADQUARTERS
ARMY AIR FIELD

370.5 # 104 (258-33)

Herington, Kansas,
13 February 1945.

SUBJECT: Movement Orders, Very Heavy Bombardment Crew Number FI-354-AL <u>21</u>, To Overseas Destination.

TO:	AC	1st Lt.	(1093)	NORMAN F. ADAMSON	0731915	AC	White
	P	2nd Lt.	(1091)	WANLESS M GOODSON JR.	0827799	"	"
	N	2nd Lt.	(1038)	JOHN J. HARNETT	0930620	"	"
	B	F/O	(1035)	FRANCIS J. GORE	T127836	"	"
	E	T/Sgt.	(737)	Ralph E. Barnett	34004096	"	"
	RO	Cpl.	(2756)	George P. Lyden	39576469	"	"
	AG	Cpl.	(580)	Jacob G. Bartscherer	42060741	"	"
	G	Cpl.	(611)	Jesse P. Colvin	34904145	"	"
	G	Pfc.	(611)	Stanley (N) Convisser	42061007	"	"
	G	Cpl.	(611)	William J. Baldwin	35779916	"	"
	G	Cpl.	(611)	Joseph O. Bischof	42003188	"	"

B-29 Project No. 98204R.   Airplane No. 42-93947   Shipment No. FI-354-AL <u>21</u>

APO No. 19053-AL <u>21</u>, c/o Postmaster, San Francisco, California.

1. You are relieved from atchd unasgd, 2AF, 274th AAF Base Unit (SS), this station, and are assigned to shipment number indicated, and will proceed via military aircraft and/or rail to Mather Field, California, or such other PAE as the CG, ATC, may direct, thence by air to the overseas destination of subject shipment. Upon arrival at PAE, control of personnel is relinquished to the CG, ATC.

2. This is a PERMANENT change of station with temporary duty enroute. Effective date of change on Morning Report is arrival date at PAE. You will not be accompanied by dependents; neither will you be joined by dependents enroute to, nor at, the PAE. You will not discuss this movement even by shipment number, except as may be necessary in the transaction of <u>OFFICIAL</u> business. You will not file safe arrival telegrams with commercial agencies while enroute to, nor at, domestic or overseas destination.

3. a. Reimbursement for officers and flight officers is authorized under provisions of WD Circulars 260 and 356, 1944.

b. Reimbursement for enlisted men is authorized under the provisions of WD Circular 260, 1944, as amended by Sec. II, WD Circular 356, 1944. EM will make use of government quarters and messing facilities if available at stop-over points. For travel by rail, monetary allowance in lieu of rations and quarters is prescribed in accordance with AR 35-4520.

c. From time of departure from the continental limits of United States until arrival at permanent overseas station, payment of per diem is authorized for a maximum of forty-five (45) days.

R E S T R I C T E D

-1-

**Transit Orders for Lt. Adamson's crew
from F/O Francis Gore**

Awards—Decorations Granted _____ *Joe Bischof* (Adamson's crew)
Date Left Continental U.S. _____ UTH. _____

DATE OF EACH COMBAT MISSION	GEOGRA- PHIC LO- CATION OF TARGET	COMBAT FLYING HOURS	NUMBER CREW CREDIT SORTIES	Cumulative Totals		INITIAL ON ENTRY	INITIAL BY CREW MEMBER
				COMBAT FLYING HOURS	CREW CREDIT SORTIES	MONTH ON DUTY	
13 Mar 45	Osaka	15:10	1	15:10	1		
16 Mar 45	Kobe	15:45	1	30:55	2		
27 Mar 45	Oita A.F.	15:20	1	46:15	3		
30 Mar 45	Tachiari	16:20	1	62:35	4		
3 Apl 45	Tachikawa	14:00	1	76:35	5		
13 Apl 45	Tokyo	13:50	1	90:25	6		
18 Apl 45	Izumi A.F.	14:30	1	104:55	7		
24 Apl 45	Hitachi	13:50	1	118:45	8		
28 Apl 45	Izumi A.F.	15:00	1	133:45	9		
19 Mar 45	Nagoya	17:20	1	151:05	10		
7 Apl 45	Tokyo	11:00	1	162:05	11		
7 May 45	WSM	14:45	1	176:50	12		
14 May 45	Nagoya	15:25	1	192:15	13		
1 Jun 45	Osaka	14:15	1	206:30	14		
5 Jun 45	Kobe	15:10	1	221:40	15		
7 Jun 45	Osaka	14:55	1	236:35	16		
10 Jun 45	Hitachi *(Tokyo)*	15:00	1	251:35	17		
19 Jun 45	Hamamatsu	14:10	1	265:45	18		
22 Jun 45	Kure	14:40	1	280:25	19		
26 Jun 45	Osaka	13:55	1	294:20	20		
28 Jun 45	Sasebo	15:05	1	309:25	21		
1 Jul 45	Kumamoto	14:00	1	323:25	22		
3 Jul 45	Kochi	13:40	1	337:05	23		
6 Jul 45	Akashi	14:00	1	351:05	24		
9 Jul 45	Sakai	13:40	1	364:45	25		
12 Jul 45	Ichinomiya	13:35	1	378:20	26		
1 Aug 45	Toyama	14:55	1	393:15	27		
5 Aug 45	Nishinomiya	14:20	1	407:35	28		
7 Aug 45	Toyokawa	13:40	1	421:15	29		
14 Aug 45	Osaka	14:05	1	435:20	30		
28 Aug 45	PWS	15:00	1	450:20	31		

Z-58 — 44-70113

Incl 1 to XXI BC Reg 35-2

R E S T R I C T E D

Lt. Norman Adamson's combat record

beneath the co-pilot's window caused by a bullet that narrowly missed his right shoulder. It was fired from a fighter that was attacking head-on. Lt. Goodson provided us with a homemade movie video-cassette that shows the crew in action and at leisure, and flying in formation with other aircraft. Though the film is very poor quality, it is thrilling to see the new *Z-58* come to life. Wanless has also sent some excellent photos of Tokyo and Yokohama in ruins, taken at low altitude (see page 8). A recent letter from him is found on page 28.

Sweet Eloise bears many other patches from combat on the skin of the aircraft, particularly in the wings, but they are hard to find. The patches are very skillfully done. It is unknown how many holes have been patched in the skin of *Sweet Eloise*. She apparently never received any major damage, and she never had any injuries to her crews.

There is a deliberate bullet hole (now covered) in each main wing tank placed there at Aberdeen in each B-29. This was to make it difficult for terrorists or any unfriendly country to restore the aircraft for use against us.

While Lt. Adamson and crew were on leave in Hawaii for two weeks, an unknown substitute crew flew two missions, on July 16 Oita, and July 19 Hitachi. Another substitute crew flew two more POW Supply Missions to China, Korea, or Japan, during the latter part of August.

In summary, *Sweet Eloise'* total targets were: Osaka (5), Tokyo (3), Yokohama, Kobe, Kure, Hamamatsu, Sasebo, Kumomoto, Kochi, Akashi, Sakai, Ichinomiya, Toyama, Nishinomiya, and Toyokawa.

Sweet Eloise completed her tour of duty with an excellent record, considerably better than the average B-29. She led a charmed life. She was an unusually fast ship. She had no aborts, and no forced landings. There were no wounded crewmen. She flew twenty-two combat missions, had five POW mercy missions, and destroyed at least three enemy aircraft. The crew received two Distinguished Flying Crosses and a Distinguished Unit Citation. This record is the result of skill and teamwork on the part of the ground crew and the crew who flew her. Plus some good luck!

Lt. Adamson and crew returned to the states on September 22, but not in *Z-58*. *Sweet Eloise* was flown back to the states by a Major Black and crew of eleven. They flew the final two POW missions.

They flew her back to the states for storage at Rapid City, SD, to Bath, MN, then to Warner Robins AFB near Macon, Georgia for long-term storage.

After the fighting was over on August 15, 1945, over 500 B-29s and 1,400 other aircraft flew in a massive victory parade on September 2 over Tokyo Bay to salute General Douglas MacArthur as he signed the peace treaty on board the battleship USS Missouri. Hundreds of ships and thousands of troops participated in the historical Victory celebration. *Sweet Eloise* did not participate in this memorable parade. At that time there were 6,000 ships en route to Japan for the scheduled invasion on November 1.

At the end of the war there were about 2,000 B-29s based on the Marianas, which needed to be scrapped or returned home for disposal. As a result of her POW missions, *Sweet Eloise* was among the last to return to the states. The return flights were complicated by the fact that Kwajalein and Johnston were tiny islands and could accommodate only a few B-29s at a time.

Recent research has revealed that the name "Sweet Eloise" was used on the side of the tail gunner's position on *Mary Anne*, B-29 # 25, on the Capt. J.W. Cox aircraft, in the 499th Bomb Group on Saipan. The co-pilot was Lt. Chester Marshall, the noted B-29 author who has published eleven B-29 books. The tail gunner, John Southerland, is credited with having shot down five Japanese fighters with two probables, making him an "Ace."

"Ace" John Southerland in *Mary Anne*
<u>**B-29 Photo Combat Diary**</u>
by Chester Marshall from his book

John held the record for all B-29 gunners during the war. His wife was named Eloise, accounting for the sign of "Sweet Eloise" painted on the side of John's tail gunner position.

Many of these veteran Superforts remained on active duty. Others were transferred to the Strategic Air Command (SAC) under Gen. LeMay, headquartered at Omaha, Nebraska.

The remainder was sent for storage to Warner Robins AFB, or Pyote, Texas, or Davis-Monthan AFB in Tucson, or Tinker AFB at Oklahoma City, joined by thousands of other surplus aircraft. Many of these veteran warbirds underwent major modification for the many different versions of this tried-and-true airplane. Many were held in reserve and moth-balled for possible future use in a changing world. The United States built over 299,000 aircraft during the war.

When Lockheed Aircraft, formerly Bell Aircraft, was activated in 1951 for the renovation of B-29s for the Korean War, the aircraft came out of storage at Pyote, Texas. Machine tools stored in the plant were removed, and Lockheed restored 175 B-29s, and started producing the B-47, a six-engine turbo-jet bomber, designed to replace the B-29. Planning began in 1952 for Lockheed's "bread and butter" plane, the Hercules C-130 transport.

"This is the weapon that will beat Japan to its knees" General "Hap" Arnold
The Souths Greatest Industry Salute You
Bell Aircraft Recruiting Display
Atlanta Fox Theater, 1944
Bill Kinney Collection

Chapter 2 ———————————

Diaries, Letters ——————

and Leaflets • Col. Ralph (Pete) Reeve • Joe Bischof, S/Sgt, DFC
• Richard Wing, S/Sgt, DFC • Wanless Goodson, 2nd Lt.
• Japanese Leaflets

Author's note: The following is a letter from **Col. Pete Reeve** who has advised me with this book. He lives in neighboring Roswell, GA

R. A. (Pete) Reeve, Col.USAF Ret.
1035 Finnsbury Lane
Roswell GA. 30075

May 25, 2000

Dear Pete,

Thank you for letting me read your manuscript. I was impressed with your knowledge of B-29s and of *B-29 44-70113* in particular. I had not realized she had such an impressive history, that she had been so close to the graveyard, and that you all have done such a magnificent job of restoration. It is a wonderful story with *Z-58* taking center stage with many curtain calls at the end.

I was one of the organizers of the 500th Bomb Group, as well as the Commander of an adjacent squadron, the 881st. I am familiar with what *Sweet Eloise* accomplished and what she endured for I flew in many of the same formations and missions as she did.

I am a Command Pilot with over 7,000 hours, and with around 1,000 hours in the B-29. I am convinced that the B-29 in its hey day was one of the most fabulous airplanes ever built.

I have attended two of your meetings of the "B-29 Superfortress Association" in Marietta and I attended the initial send-off of the nose section for rehab as well as the big dedication, meeting old Saipan buddies on both occasions. I have visited *Sweet Eloise* several times in the past three years, and each time I have out-of-town visitors, they get the big tour. I will be able to give them an even better tour with the help of your book.

We all appreciate what you folks in Marietta have done in preserving our legacy. Those of us who "have been there and done that" - and survived - say "Thank you." We remember the 3,000 or more of our buddies whose luck ran out, and who never made it back.

You have taken the story of an old warrior, woven in some sub-plots, and made an interesting story out of it – a story needing to be told many times. Thank you again. Good luck on your book.
Sincerely,
Pete

Diary of S/Sgt Richard C. Wing, DFC

Author's note: S/Sgt Wing was a gunner on Capt. Clinkscales crew. They flew in several different B-29s, including *20th Century Sweetheart, Supine Sue, Ancient Mariner,* six missions were in *Sweet Eloise.* Written on board during each mission on flight-log note-book.

Summary of Raids

12/20/44 Nagoya Hours 13:55. Dock area
1st was weather strike. Bombed by radar. Picked up by search lights with intense flak. Encountered two night fighters. Sustained no damage. Bomb results unobserved.

12/27/44 Tokyo Hours 11.05 Fish
2nd mission we had to turn back because of blown cylinder head. Raid was successful. We lost one B-29 over the target and five others ditched.

1/3/45 Nagoya 13:45 Aircraft Factory
3rd mission was an incendiary attack. Results were good. Flak was very heavy & accurate. Encountered 200 enemy fighters, 63 were shot down. Lost 3 B-29 over target and two others ditched on the return home.

1/27/45 Tokyo Hours 14:25 Dock area
4th mission was largest group of B-29 to ever attack. Bombing was excellent. Flak and fighters were intense. Shot down my first Jap fighter, a Tony. Tail gunner Sgt. Glockner and blister gunner Sgt. Chambers also shot one each down. Five B-29 were shot down and two ditched on return to base. Our plane was damaged by flak and fighters.

2/10/45 Oita Hours 15:15 Aircraft Factory
5th mission was an aircraft factory. Flak and fighters were light. *75% damage on target. Three B-29 ditched on return to base.

2/19/45 Tokyo Hours 14:50 Engine Plant.
6th mission encountered bad weather. Dock area bombed by radar. Results unobserved. Flak and fighters were intense. We lost nine B-29 over target. "Shot down," No damage sustained to our ship.

2/25/45 Tokyo Hours 14:35 City Area
7th mission was to destroy Tokyo by fire. 200 B-29 took part. Radar bombing, good results. Two B-29 collided and went down in flames. Weather was bad. Light flak and few fighters.
246 square blocks destroyed.

3/1/45 Search Mission Hours 13:55 B-29 crew
8th mission was to search ocean for missing 29 crew. Crew was found and picked up by US Navy destroyer.

3/10/45 Tokyo Hours 14:50 City Area
9th mission was same as the 7th with 300 B-29 attacking. This was first low altitude bombing mission flown at night at 7,000 ft. Guns and ammo were removed. Results were excellent. Flak was intense. Smoke rose to 20,000 ft. 15 square miles of Tokyo were burned out with a loss to the Japs of 200,000 lives. Three B-29 were shot down over target. "Rough"

3/13/45 Osaka Hours 15:00 City Area
10th mission was low level attack on Osaka with incendiary bombs. 300 B-29 took part. About ten square miles of city was burned out. Flak was heavy. Eight B-29 were shot down. No damage to us.

3/16/45 Kobe Hours 14:50 City Area
11th mission was to burn out Kobe. Attack at 5 AM at low altitude, 6,000 ft. Bombing was excellent. Flak was intense, few fighters. One B-29 shot down over target. Fire could be seen for 80 miles back out from coast.

3/24/45 Super Dumbo Hours 16:50 Search and Rescue.
12th mission we flew Dumbo in search of ditched planes returning from mission. Searched for one with no results. Plane was thought to have blown up and sank with no survivors.

3/27/45 Kyushu (Oita) Hours 15:10 Airfield.
13th mission we flew to Oita to destroy an airfield. Our first to this part of Japan. Went in at 15,000, daylight. Flak was meager, no fighters attacked. Results of bomb hits were good One ship ditched on way back and three landed at Iwo Jima.

4/1/45 Tokyo Hours 14:10 Mishashino Aircraft Factory.
14th mission was to destroy Mishashino Aircraft factory. Heavy flak and search lights. Bomb results were good. No damage to our ship, but one was seen to have been shot down. Loss was said to be seven planes.

4/7/45 Tokyo Mishashino Aircraft Factory.
15th mission was to bomb Mishashino Aircraft Factory. Daylight raid at 15,000 ft. Intense, accurate, and very heavy flak. P-51 from Iwo Jima escorted us in. About 300 Jap fighters attacked. Many were shot down. Our crew has definite kill of five. I shot two more down and both pilots bailed out. This was our squadron's roughest mission to date. We lost one plane and crew over target. All planes sustained large flak holes and some had to land at Iwo Jima on return home. We dropped out of formation over target to protect crippled plane behind which Jap fighters were attacking. We fought off fighters and stayed with the plane till it reached Iwo Jima and landed. Bomb results were good. Our plane had flak holes and antenna shot out.

Author's note. Capt. Clinkscales' aircraft aided this crippled plane being piloted by Lt. Norman Adamson, both of whom are future pilots of *Sweet Eloise.* See "So You Wanta Ram," page 11 and 13.

4/11/45 Super Dumbo Hours 17:05 Search and Rescue
16th mission was a Dumbo to search for downed aircraft on return from Tokyo. Worked with destroyer, in search of P-51 pilot downed at sea north of Iwo Jima. Sea marker seen but pilot wasn't found.

4/12/45 Tokyo Hours 13:55 Incendiary Raid
17th mission was another incendiary raid on Tokyo. We flew in at 7,400 ft. Searchlights and flak were heavy. Some 50 plus Jap fighters were seen. We were lucky as no Jap searchlights picked us up and no fighters attacked. The plane on our right was shot down by flak. Our squadron suffered no losses but two B-29 were lost over the target and others crippled. Bomb results were good and fires were widespread. You could smell the smoke and Tokyo was lit up like day. Big explosions were seen miles back out to sea.

4/15/45 Super Dumbo Hours 13:50 Search
18th mission was another Dumbo mission to help cripples and guide surface vessels to ditched aircraft. On take-off we had a narrow escape & had to use another plane, *Gen 0'Donnell* to complete our mission. Our first plane blew a cylinder head & #4 engine was afire. We were just off the runway with an airspeed of 130 mph. The pilot cut throttles and thanks to God the brakes held. We stopped in time & fire dept. put out the blaze. We had no emergencies to take care of, although some planes did ditch close to a destroyer so we were of no assistance. Returned to base.

4/18/45 Nagoya. Hours 13:20 Recon Radar Photo
19th mission was to take radar photos over Nagoya. We carried no bombs. Went in at 14,000 ft about 3:30 in the morning - 10/l0s cloud coverage. Picked up two searchlights & no flak. Easy mission for a change.

4/22/45 Okinawa Hours 12:55 Escort P-38 to Photo
20th mission was to escort 4 P-38 to Okinawa. Two B-29 took part, us being one. We had no opposition as the Navy was really pouring it on. Seen fighting on the ground below. Also large Navy task forces (U.S.). I took a few pictures.

4/26/45 Kyushu Hours 14:05 Sacki Airfield
21st mission was to attack Sacki Airfield at Kyushu, Japan. Bombed thru overcast from 21,000 ft altitude. Results unobserved. No flak and no fighters.

5/15/45 Honshu Hours 13:00 Naval Base
22nd mission was to a naval base on Honshu. It was a clear day & seen very much of Japan as we were over land about 1 ½ hours. We dropped four 2,000 lb bombs from 25,500 ft. The results were excellent, targets destroyed. There were 11 planes in our sqdn, some received flak damage but all returned. No fighters encountered but flak was intense and accurate over the target. Due to bad weather on the return, we landed at Iwo Jima and spent the night and part of today. Still fighting Japs there so we didn't get around much. Went thru cemetery and found K.G. Wing.

5/14/45 Nagoya Hours 15:05 City of Nagoya
23rd mission was incendiary raid on the city of Nagoya. Daylight raid at 16,000 ft. We encountered meager flak and six fighters on the way to the target. No damage done to our squadron. Bomb results were very good & smoke was up to 15,000 ft.

5/18/45 Honshu Hours 14:30 Tachikawa
24th mission was to fly to Iwo Jima- stay overnight and then escort 90 fighters to the target of Tachikawa. Due to bad weather, the B-29s bombed Homamatsu, with good results & we brought the fighters back to Iwo. We were out from base 1 1/2 hr. & lost #3 engine, had to return. Flew # 56 to Iwo and had a flat tire next morning. What a mission. Glad to be back.

5/23/45 Tokyo Hours 14:15 Incendiary Raid
Author's note (First of six missions in *Sweet Eloise*)
25th mission was to burn out southern Tokyo. This was a night mission and bombed from 11,800 ft. Some 600 B-29 took part. Our losses were six planes plus over the target & more ditched on return to base. We were lucky as we seen no flak or fighters and missed all searchlights. Returned safely. Results were estimated at 30 sq. mi. burned out. "What a fire!"

5/25/45 Tokyo Hours 14:50 Incendiary Raid
26th missions was the same as the 25th, only further north in Tokyo. Fighters and flak were heavy. We sustained no damage, why I don't know as we were caught in searchlights & shot at for ten minutes. One of our toughest missions. Some 25 plus B-29 were shot down.

5/29/45 Yokohama Hours 13:55 Incendiary Raid
27th mission was a daylight strike to Yokohama. We went in at 18,000 ft, dropping 185 100 lb bombs. This was an incendiary raid. 450 B-29s took part. Flak was heavy and had few fighters. Smoke rose up to our altitude. The town was really burning. We had 12 bombs hang up, and worked on the way back to release them. We were a little worried but got them out OK with only two holes in the bomb bay doors. We lost two B29s and suffered one flak hole. Returned safely. (5 more to go?)

6/1/45 Osaka Hours 14:30 Incendiary Raid
28th mission was another daylight incendiary raid on the city of Osaka. We went in at 19,000 ft. The flak was heavy and accurate. We lost one '29 out of our formation. Bomb hits were good & smoke rose to 30,000 ft. We weren't hit & returned safely to base. "Only four more to go!"

6/5/45 Kobe Hours 14:55 Incendiary Raid.
29th mission was to burn the city of Kobe, which was done. Some 450 B-29 took part. Flak was intense and accurate with some fighters. Seen one 29 go down in flames. We weren't damaged. Three more to go.

6/7/45 Osaka Hours 14:30 Incendiary Raid.
Author's note (Last of six missions in *Sweet Eloise.* Transferred to Major Black's crew.)

30th mission was another burn raid on Osaka, the second within one week. Bombed thru overcast. Very little flak and no fighters. This was an easy mission. Two more to go, thank God. We're all pretty tired and homesick.

7/1/45 Kumomoto Hours 14:00 Incendiary Raid 31st mission was on engine factory, north of Tokyo. We carried seven 2,000 lb bombs and dropped them from an altitude of 20,000ft. Results were excellent & the factory was missing after the smoke cleared. Flak was meager and inaccurate and no fighters were seen. Returned safely. Thought we had one more to fly, but now it's four more. With good luck, I should be finished, completed, by this month. This was a secondary target as our old favorite #357 was clouded over, which made us all happy!!

32nd mission was an incendiary raid on Osaka. The weather was so bad we had to go over the city and drop individually. No flak and no fighters.

33rd mission was a night incendiary raid on the city of Kumomoto, Japan. We bombed from 12,400 ft, thru an overcast. The fires were big. Flak was meager & no fighters were seen. Four cities were burned this night, with the biggest force of B-29s to ever attack Japan. Some 600 planes took part. This mission is it. Homeward Ho!

THE END

S/Sgt. Joe Bischof's, DFC. Combat Diary

Author's note: S/Sgt. Joe Bischof, from New Bergen, N.J. was the tail gunner on Lt. Adamson's crew. The co-pilot was Lt. Wanless Goodson whose letter appears on page 28. They flew 13 combat missions over Japan in Z-53, *Ancient Mariner.* They were transferred to Z-58, *Sweet Eloise,* and flew 17 more missions in her, plus one POW drop, which was on Mariettan Bill Price's POW camp outside Tokyo. Enclosed is his official flight log and crew photo. He flew 450 hours and 20 minutes on 31 missions. Most of the missions lasted 14 to 16 hours. He recorded some of his thoughts and experiences, a little quiz, some notes and excerpts from his diary. This was on five different aircraft, but with the same crew. The diary is quoted from Alton Evans book, B-29, 44-70113, pages 15 to 42.

MY COMBAT MEMOIRS
Joseph Bischof, S/Sgt. DFC
To Jerry Bartscherer I dedicate my combat memoirs. It all took place in the Pacific Theater of War in 1945. I will always remember you as my closet buddy and confidant. We went into battle and survived thirty combat missions over Japan. Your cool, contemplating manner is part of the reason why we survived. That one act of bravery going into the aft bomb bay with our bombardier, and working to release a hung bomb made you number one in my book of memories. Many times when I thought we would never survive with so many missions yet to accomplish, you gave me strength to see on to the final victory. To you, Jerry, I dedicate the following historical notes.

This is the story of a B-29 bomber crew out of Saipan that flew thirty bombing missions over Japan and never aborted one. Our crew was a closely knit group that was formed in Grand Island, Nebraska. From Grand Island on to Pyote, Texas for crew training. We had many flying hours over Texas, Oklahoma, Kansas, Nebraska, and other mid-western states. Finally, we made a round trip to Cuba and back to Galveston, Texas.

By early February 1945, we were combat ready. We were part of the 20th Air Force, (Army Air Corps). Ultimately we were assigned to Saipan to begin the bombing of the Japanese Empire.
73rd Bomb Wing...C.O. Gen. Emmett "Rosie" O'Donnell
500th Bomb Group ...............C.O. Col. John E. Dougherty
883rd Bomb Squadron ...C.O. Lt. Col. William McDowell

The following is a day-by-day account of our activities as researched from my letters that were written home almost every day. I am thankful that my dear mother saved all my letters.
Feb. 12, 1945 Arrived in Harrington, Kansas to pick up a new B-29.
Feb. 13 Crew picture was taken. Each received two copies.
Feb. 17 Arrived in California.
Feb. 20 Somewhere in the Pacific. (From Calif we headed for Hawaii, Johnston Island, Kwajalein and finally Saipan. The airplane made excellent time. The A/C became *Z-55* and was assigned to an older crew.
Feb. 26 Started on my beer cooler. Week's ration was 6 bottles, 2 cold and 4 warm.
Feb. 29 Stopped smoking cigars. Sold $3.50 box to Jerry for $2.00
Mar. 2, 1945 Saw "Fighting Lady" movie. Had 45 cal. inspection.
Mar. 3 We sit on bomb stools at open air movie.

Played horseshoes with George Lydon. We practice formation bombing on Jap island.

Mar. 4 Laundry is sent out every 10 days. Attended Mass at 12:00 noon. Gunnery class to start soon.

Mar. 6 Adamson and Barnett go on Tokyo mission. No losses. Have not received any mail yet.

Mar. 7 Jerry and I build a card table.

Mar. 8 Received first mail.

Mar. 9 Used Colonel's shower and did not get caught.

Mar. 10 Getting good at pinochle.

Mar. 13 Got paid. Bill paid back $30.00.

Mar. 14 Bombed Osaka. No resistance.

Mar. 16 Bombed Kobe. Big fire even before we got there. Flew through thermal - tossed up 1,000 ft and down just as fast. One bomb stuck in shackles - put hole in bomb bay door, later released over ocean.

Mar. 19 Weather mission to China coast. Bombed Nagoya on way back. Landed on Iwo Jima for fuel - then returned to Guam with weather info. It was a long trip - on my 20th birthday.

Mar. 26 Bill, Japer and I visit the Navy. Ralph makes Master Sgt.

Mar. 27 Bombed Oita A/B. Only tail guns were loaded.

Mar. 29 Installed 5 gal. oxygen tank for beer cooler.

Mar. 30 Bombed Tachiari A/B. Some flac. Tail guns and upper aft turret loaded.

Apr. 4, 1945 Bombed Tachikawa at night. Almost fired on B-29 on our tail. He was slow answering on IFF.

Apr. 7 Bombed Tokyo, Z-48, daylight, 12,000 ft. We had 7,500 rounds. The flak was heavy. We were attacked by many Tony's and Tojo's - over 50 fighter attacks. We damaged 8 and shot down 3 or 4. More details are on another document. This was one of our most dangerous missions. If Commander Clinkscales had not dropped back to help us fight off enemy fighters I'm certain we would have been lost. We had 2 engines out and over 500 flak holes in the ship. We made a forced landing on Iwo Jima. It was 3 days later that we finally returned to Saipan.

Author's note See page 11 and 13

Apr. 14 Returned from Tokyo mission. President Roosevelt dies.

Apr. 15 Attended memorial service for President, in class A uniform.

Apr. 17 We have a new gunner on our crew replacing Convissor. His name is James Emerson.

Apr. 18 Bombed Izumi A/B.

Apr. 20 Fired my M-1 Gerand today. We heard that the war in Europe will be over soon.

Apr. 22 Finished building my writing table. Guys in our Quonset are honest. I leave my wallet on my bed when I go on a mission.

Apr. 23 Started writing to Jerry's niece Patricia Hogan. She saw me in a photo that Jerry sent home.

Apr. 24 Bombed Hitachi south of Tokyo. I think I damaged a Jap fighter. Mt. Fuji is beautiful - covered with snow.

Apr. 25 Saw "A Tree Grows in Brooklyn." Cleaned guns on ship.

Apr. 27 Saw Danny Kay imitator. Building porch on barracks.

Apr. 29 Bombed Izumi. Easy mission. Low on fuel. Landed on Iwo Jima.

May 1, 1945 Got Sgt. Stripes.

May 2 Heard rumor that Hitler was dead.

May 4 Have not flown a mission in a week.

May 6 Gore and I toured Saipan. A C-47 sprayed the island at low altitude.

May 7 Saw Navy band in stage show. Jackie Cooper was the drummer. Dennis Day sang.

May 9 12th mission. No details.

May 15 Bombed Nagoya for our 13th mission. It was an incendiary raid. I don't like daylight burn raids because you can't see the flames as you can at night. Smoke rose to 15,000 ft - saw a few fighters - damaged one.

May 18 Very little flying in May. Missions getting easier. Food getting better.

May 20 I explained in this letter why residential areas had to be bombed. Almost every Jap home is a factory. Cottage industry as we call it today.

May 21 Someone got a CFC blister and we made it into a fish tank. We netted some tropical fish and placed the tank in the shade.

May 22 My mother sees her first B-29, probably in a news reel.

May 24 All the tropical fish died. No missions lately because we are going to school for two weeks. We were told that rest camp is now in Australia.

May 27 We now have the Air Medal with one cluster.

May 28 We had to remove all nose art from B-29s.

Before getting a steady B-29, I had flown in *Three Feathers, Twentieth Century Sweetheart, Fancy Detail, Tail Wind,* and several others. We did not name our ship yet. I like *Old Ironsides.*

June 1, 1945 Our 14th mission was to Osaka. It was incendiary. Smoke plumed high above our altitude. Gore shot down a Betty. It disintegrated as it passed below our A/C on a 12 o'clock attack. Father Hickey blessed our ship before take off. F/O Francis Gore is confident that he shot down the fighter, but he did not claim credit for it.

June 5 Returned from a raid on Kobe. I hope the Japs surrender soon. Smoke was visible 100 miles out to sea. This was our 15th mission (see page 5).

June 7 Returned from fire raid on Osaka. Saw no enemy A/C (Aircraft) on this raid. Several puffs of flak. Easy mission.

June 8 I think we will be going to rest camp soon.

June 10 Spent most of Sunday in the air. Bombed Hitachi, an industrial section of Tokyo. Have not had a fighter come in on me since April 7th raid on Tokyo.

June 11 Attended a religious mission followed by Benediction.

June 12 Father Dolan gives sermon in the rain.

June 15 I send a picture home of *Three Feathers* - the A/C we flew on the April 7th mission.

June 17 We still can't reveal what island we are on.

June 18 We have been assigned a new plane. It is about 4 combat weeks old.

Author's note First of 17 missions in *Sweet Eloise.*

June 19 Bombed Hamamatsu. Had so many search lights on us it was like daylight in the tail.

June 22 Returned from Kure raid - our 19th mission. This was a daylight demolition raid on Japan's largest naval repair station and manufacturer of large anti-aircraft guns. Much flak - black, yellow, purple - so much flak I could hardly see the squadron behind us.

June 24 I hear Mass 2-3 times a week. A little house beside the chapel has the Blessed Sacrament exposed 24 hours a day. I helped paint the house for Father Mickey. We had beer and soda. I made a visit before and after each mission.

June 26 Bombed Osaka Navy Base - demolition - daylight. No flak, no enemy fighters. Had P-51 escort from Iwo Jima. Heard Jap broadcast on Okinawa loss "Our Imperial troops gained much experience - We have not lost the battle." Some propaganda.

June 27 Cleaned windows in tail, cleaned floor, etc. Got new lightweight electric flying suit. Very comfortable.

June 29 We fire-bombed Sasebo last night. On the way out saw another city on fire by another Wing.

June 30 Received Asiatic Pacific Campaign ribbon with 2 battle stars. I also have the DFC on the way.

July 2, 1945 Burn raid on Kumomoto. No flak. No enemy fighters. Our 22nd mission.

July 3 Burned Kochi. Saw two cities burning at once. Flew through smoke - like a Roller coaster ride. No flak or fighters.

July 5 Rest leave soon - maybe Australia.

July 7 Returned from a night raid on Akashi near Kobe. Had a K-20 camera in tail. Took pictures of bomb results. No fighters or flak. Our service club opened. All had 2 free beers.

July 9 Spent most of day in ocean riding my air mattress.

July 10 Completed night burn raid on Sakai adjacent to Osaka. We were 2nd to drop bombs. Many lights, no flak. Watched the city burn brighter and brighter as we left. Saw another city burning a few miles away. It was the most furious fire I have ever seen. Have not seen enemy fighters in last 10 missions. When our Bomb Group started they flew about every 10 days. Some took 3 months to complete 7 missions. When we came, we averaged one every 3 days. We are up with the original crews and ahead of one of them. Jerry and I receive Viaticum before each mission. Fr. Gaspar and Fr. Hickey are up on the runway before each flight giving blessings. I may make S/Sgt next month.

July 13 Completed 26th mission - night burn raid on Ichinomiya. First to return to Saipan. - our new ship is very fast. No more missions for at least 3 weeks. Leaving for Oahu tomorrow. Was disappointed it is not Australia.

July 16 Left Saipan for Hawaii on July 14 in C-54 transport. Landed at Kwajalein, Johnston, and finally Hickam Field. Had choice to stay at Hickam or Waikiki. We chose Waikiki. Jerry and I attended Mass at Hickam at 12 noon and then went to Waikiki. Jerry and I take bicycle tour of island.

July 18 Swim every day. Food is great. Lt. Goodson remained at Hickam, and Barnett is at Holiday House.

July 25 Leaving today for Saipan. I read where Japs are taking a beating. I am sure missions will be less strenuous. At a dance here I met 2 girls -

one from Union City and one from Weehawken. Great food here - steak and chicken every day. Back on Saipan we will have goat meat almost every day.

July 30 Back on Saipan. On the last mission no A/C were lost from all bases in the Marianas. My MOS # has been changed to 612. I should get S/Sgt. in a few weeks.

July 31 Sent leaflets home. Thousands were dropped on Japan urging them to surrender. Jose Sablam gave me a hand-carved Knaka doll.

Aug. 2, 1945 Participated in an 800 plane mission. Our Wing hit Toyama. We announced the raid in advance by dropping leaflets so the Japs knew we were coming. No flak or fighters. Fire very big - I guess the whole city was destroyed. Original crews starting to go home. New crew arrives from Pyote.

Aug. 3 One original crew returning to the USA by boat. Only one crew in squadron has more missions than us. Hoping Lt. Adamson gets Capt. rank soon. Jim Emerson has 32 missions. He came from Maj. Black's crew with 5 missions.

Aug. 4 Saw best show yet. Larry Adler's brother played the harmonica. A Seabee tried to remove our "rock" with a bulldozer and failed. My leaflet was removed from letter by censor.

Aug. 6 Mission #28 was a night burn raid on Nishinomiya. Lt. Adamson was Promoted to Capt. He passed out cigars. He was a First Lt. over 2 years.

Aug. 6 ATOMIC BOMB ON HIROSHIMA (not in letter because it was not known on drop date).

Aug. 7 Mission #29 was Toyokawa. We have flown 30 hours in the last 48 hours.

Aug. 8 We are permitted to mention Saipan, Tinian, and Guam. All day long we hear how devastating the new bomb is. It sounds like the bomb to end civilization.

Aug. 10 Had guard duty - learned to drive a jeep.

Aug. 11 Capt. Adamson took our 45s to avoid any dangerous celebration at wars end. Did not get to sleep until 3:00 AM Our crew is second highest in missions flown.

Aug. 13 Made S/Sgt. We are on alert for another mission. We got a truck and took a trip to the highest point on the island.

Aug. 15 Mission #30. Demolition on Osaka. Japs were pleading they had surrendered. Unofficial - We dropped the bombs as ordered. Then the flak begins to appear.

Aug. 16 Beer ration doubles up to 12 bottles per week. I explain the job of a lead ship. Our *Z-58* was very fast and we usually returned first. Once my guns jammed over Tokyo. All I could do was point them at Jap fighters and try to scare them off. It worked sometimes.

Aug. 19 Our ship has a PW on it. I believe it will be used to pick up POWs in Manchuria. It will be a 9 hour flight from Saipan. We can enjoy the island now. No more getting up at midnight for a daylight raid.

Aug. 21 Jerry gets a truck and he lets me drive. Had to relocate to another Quonset. Newly arrived officers took ours. They complained about the war's end and having to go home without any medals.

Aug. 22 Performed a practice bomb run on an uninhabited island. It had a smoking volcano in the center where the bombs hit. With all the flak damage we had on many missions, no one was ever wounded. A small fragment did graze Bill's forehead.

Aug. 24 Attended Mass by Archbishop Spellman. Our General, Emmett "Rosie" O'Donnell attended and received Holy Communion.

Aug. 27 Jim Emerson leaves for Guam to return home - gets bumped by a Colonel. He had given me a box of tools and a P-38 airplane kit. I wanted him to take the stuff back but he refused. What a great guy.

Aug. 28 Nigota POW run. We just returned from Iwo Jima from a supply mission 150 miles north of Tokyo. A pleasure to fly low over Tokyo and not get fired at. At times I think we were about 200 ft. altitude. Parachute supplies were dropped from 500 ft. Saw people in town - carts, bicycles, people on rooftops. One crew spots train stopping, crew gets out and throws coal at B-29.

Author's note Last of 17 missions in *Sweet Eloise*.

Aug. 31 Building the P-39.

Sept. 2, 1945 Had to say Goodbye to George. Just realized what a great friend he was.

Sept. 5, Jim left 2 days ago. Going home by boat. Got the order for my 4th Air Medal today. I also think we will get the Presidential Unit Citation.

Sept. 5 Hurricane on Saipan. Tore all tarpaper off mess hall roof. Short circuit. Everything gives you a shock. George is in California.

Sept. 7 One squadron leaves every night. Tinian flies one night and Saipan the next. Kwajalein can't hold many B-29s.

Sept. 9 This letter tells of my hatred for the Japs. Col. King who was shot down and captured before I arrived in Saipan, returned from POW camp and spoke to us. He said B-29ers were treated as criminals and given 1/2 rations. Of the 1,000 B-29 crew members captured only 170 are alive and most of them will need two years recuperation. I tell about that terrible Tokyo mission (April 7) when No. 1 engine caught fire. After forced landing on Iwo Jima, it seemed the whole island was out looking at our ship and wondering how it ever made it back.

Sept. 10 This letter tells that we returned from Iwo Jima on April 9th. We counseled with intelligence officer and re-counted to 80 enemy fighter attacks on our aircraft. This letter has much detail on our stay at Iwo.

Sept. 11 An air parade in planned. A banana fell through my model P-39's wing.

Sept. 11 The war is over but the food situation is very bad. Have not had fresh meat in a month. Goat meat twice a day. It killed my taste for lamb.

Sept. 14 73rd Bomb Wing going home in 10 days.

Sept. 17 Leaving Saipan in 7 days. Had hopes of going in *Z-58*.

Sept. 17 Sunday Sept 22 is date set for 883rd Squadron to leave Saipan. Get home for sauer braten and kadouffel claise.

Sept. 18 Return delayed again. Capt. Adamson thinks we go in *Z-58*. All plans rapidly changing.

Sept. 21 We have been put in for another DFC. Supply mission counts as #31.

Sept. 25 First planes leave Saipan tomorrow. This morning I received my medals in a formation on the field, I received Air Medal with 3 Clusters and the DFC(Distinguished Flying Cross). I'm sure now that we will receive 2nd DFC. (One of two crews to receive 2 DFCs).

Sept. 28 Received 2nd DFC and 2nd Battle Star on Asiatic Pacific Medal. Flew to Tinian to pick up bomb bay platforms to carry more passengers. THANKS BE TO GOD!!!!!

Recollections by Joe Bischof
Which Mission Were We On When . . .
1. We force landed on Iwo Jima the first time?
2. Were we the first B-29 to use Iwo Jima in emergency?
3. We force landed on Iwo Jima the second time?
4. We counted over five hundred flak holes on our first aircraft?

5. We lost one bomb bay door due to flak damage?
6. An enemy shell lodged in the aircraft beside Lt. Goodson's right arm?
7. When Capt. Clinkscales dropped back when we could not keep up with the formation and fought off approximately thirty enemy fighters as they continued their attack out into the sea? (see page 13)
8. We were attacked by Kamikaze fighters while flying in formation and one aircraft flew through No. 2 propeller at ten o'clock and barely missed our fin and rudder.
9. A bomb hung up on a bomb bay rack. F/O Gore and C.F.C. Jerry Bartscherer had to go in and manually release it?
10. We lost No. 3 engine because of a twenty mm. bullet hole in the blade and No. 2 engine front oil sump was hit and caught fire (flames were streaming back past my tail escape window). Didn't T/Sgt. Barnett direct all the C02 to No. 2 engine that was already feathered, however the fire was still burning? We were preparing to bail out over Japan but Lt. Adamson dove the aircraft at about 45 degree angle with the cowl flaps open and blew the fire out?
11. We returned to Saipan, two engines out, low on fuel, skimming over the water, had to gain altitude to land, and right after landing the remaining two engines stopped -out of fuel. We had to be towed into the hardstand?
12. F/O Gore nailed a Japanese "Betty" as it came in at twelve o'clock. I was told it would pass under us so be ready to fire. I aimed my guns accordingly but all I saw appear was a big ball of flame and many flying pieces?
13. Lt. King and crew were lost off our right side. All engines were going - no one bailed out. Flight deck must have been hit?

Remember When? by Joe Bischof
A. The last raid, Osaka, blue sky-white puffy clouds-no flak- the Japanese radio pleading "Don't drop bombs, war is over. We surrender," but we dropped as ordered and then had to fly through a heavy flac barrage-not just the familiar black puffs- but blue, red, and orange bursts?
B. When we hit the fifty-five gallon drum with our No. 2 propeller while taxiing from our hardstand on our first combat mission?

C. When Major Gay came to our Quonset (after surrender) and was asking for scanners for a flight to Guam to show his Flight Surgeon friend of his how a B-29 flies? All were killed in a cart-wheel crash on Guam. During touch and go short field landing he forgot to raise full flaps and lost the aircraft.

D. The complete destruction of Sasebo - we dropped fire bombs on the center of town from 7,000 ft then slowly gained altitude to 18,000 ft. while photographing the action below? The following B-29s perimeter-bombed the outside ring of flames gradually engulfing the whole city. I was told years later by a POW that we had totally destroyed a rice warehouse with the town's year supply of rice and starvation got even worse after that.

E. The night burn mission when we flew through the thermals over the already burning city and were abruptly forced up a few thousand feet and came back down just as abruptly? I remember denting my helmet as my head hit my ceiling-mounted oxygen bottle.

F. When returning at high altitude and my oxygen bottle was not in my tail compartment? I tried to make it forward without it. Lucky for me, you guys were waiting for me, to administer oxygen or I would not be here today.

G. After "Bombs Away" on a daylight mission we ran into heavy fighter attacks and Lt. Adamson took evasive action and flew the B-29 like a fighter? In one of the steep turns and banks the tail skidded so violently (vibrating from left to right) that my head (in helmet) was hitting the side windows with such force that I had to put my arms up to fill in the sides and form a cushion for my head.

H. I inadvertently inflated one chamber of my "Mae West?" Boy, did that ever tighten up my parachute harness. After that incident I kept a knife handy in case my seat raft ever inflated. I'm sure that if the raft ever inflated in the tail compartment it would kill the tail gunner.

I. On the night missions Lt. Adamson would ask the crew, "Lights On or Off?" and we would always say "On!" The Japanese could always find us with their radar lights anyway so with all our lights on there would be less chance for a midair collision. I think that is how Lou Doric and his crew were lost.

J. I remember having so many spotlights on us you could read a newspaper on the floor? Then George Lyden would drop the aluminum foil rolls and all lights would descend which proved they were radar activated.

K. After I fired on a P-51 who chased a Japanese fighter through our formation, the P-51 then zoomed below us and cut our trailing antenna wire? Lucky for him he didn't hit the weight.

L. Even though I was first in my class on enemy aircraft identification, in action it was difficult to identify as a dogfight was taking place through our formation? The action was so fast that at times it was best not to fire at all so as not to hit another friendly aircraft.

M. The time I held my fire as a Japanese fighter came in at six o'clock level and you guys were yelling "He's yours, Joe. Fire! Fire!" Being low on ammo I held my fire carefully keeping my dot on the pilot's head. When I saw his leading edge guns flame, I fired one short burst. One less Japanese pilot in a perfectly good aircraft with a broken windshield.

N. The night while flying to Japan we were followed by an unidentified aircraft, we challenged him on IFF (Identification Friend or Foe) and he did not respond? I was ordered to turn all switches on and to frame the aircraft (its red and green wing tip lights were on) in my sights. At this time all that remained for me to do was to press my thumb trigger. Finally he responded on IFF. He was one of us - a Black Widow fighter-bomber probably up from Iwo Jima.

O. The time we force-landed on Iwo Jima with our first B-29? We were told to "Land in the ocean, runway not secure, we'll send out a taxi to pickup survivors." Lt. Adamson answers - I see vehicles on that strip - get them off - we are coming in. We then landed against Marine orders with no brakes. However, the bomb holes helped stop us before running into Mt. Surabachi where the Marines were fighting.

THE END

Author's note: The flights from May 23 to June 14 were made in *Sweet Eloise* by Capt. Clinkscales and crew, including gunner S/Sgt Dick Wing. The flights from June 19 to August 28 were made by Lt. Adamson and crew, including 2nd Lt. Wanless Goodson and tail gunner Sgt. Joe Bischof, in *Sweet Eloise*.

Author's note: The following is a recent letter from **Lt. Wanless Goodson**, who was the co-pilot for Lt. Adamson on most of Joe Bischof's missions. Wanless narrowly escaped injury to his right shoulder from a Japanese bullet. The patch in the aluminum is still visible. Lt. Goodson attended our first dedication in 1994. Wanless has sent me photos that he took of the ruins of Tokyo and Yokohama from low altitude which he took after the armistice when no one was shooting at his plane. He also took "home movies" of *Z-58* in action. After the war he took photos of the "Pile of Junk" at Florence, S.C.

Wanless Goodson, 2nd Lt.
P.O. Box 858
1481 Stewart Street
Welch, W.V. 24801

December 13, 1999
Dr. Pete Inglis
80 Lindley Ave.
Marietta, GA 30064

Dear Pete:

I want to thank you for your willingness to send us these communications about our favorite airplane. I especially appreciate all the effort you and your group have done to make this great airplane a memorial to the great effort made by so many people in such a short time, to build a piece of equipment like this, and the many people who flew and maintained it.

From my position there are very few people who were in this great war, who can look back and see a piece of equipment like this which they used, still intact. I remember my last visit to Florence, and how despondent I was at the time. I told my wife then that I did not want to see it again if it stayed in that condition. I feel that it is something of a miracle that your group came along and did such an outstanding job in restoring it. My first visit with it was when we learned about it in Maryland.

A friend of mine who flies and I made a trip to Maryland to see it, and on several occasions I saw it in South Carolina.

I have read with great interest the copy of your book that you sent me. (Al Evan's book) I want to compliment you on your effort. I think it is outstanding, and I certainly want a copy when it is available. It helped me to remember some of the experiences and events in my history. One bit of the experiences I had I remember as different from your account. I went on the route from Hawaii and Saipan 4 times. Once when we flew a plane over and the return to the U.S., and the round trip on the rest leave. The stop over was on Kwajalien Atoll each time. I remember a sign on the operations office. "Kwajalien A Toll. No Women A Toll, No Beer A Toll, Nothing A Toll." Kwajalien was a tiny island, with the runway just about filling it up. I think is has been a storage place since the war for some dangerous items, such as Chemical Bombs.

Thank you again for the time you have given me and especially the pictures. I am going to have the one of the airplane enlarged.

I hope the enclosed check is enough to pay for the book, shipping costs and picture cost. If I know where I can buy a copy of the Greenland B-29, I would certainly like to know.
With Kindest Regards
Wanless Goodson

Japanese Leaflets

Author's note: The B-29s, including *Sweet Eloise,* dropped thousands of leaflets on the Japanese cities several days prior to bombing them, urging citizens to evacuate, and urging them to replace their militaristic leadership. These leaflets had a chilling effect on the populace and were very demoralizing. They were effective. The following five leaflets were obtained from former crewman, F/O Francis Gore, navigator, from Jamestown, N.Y. They have been translated by Dr. Koji Yoda, PhD, a former resident of Japan, who now lives in Marietta with his family.

Face of Leaflet

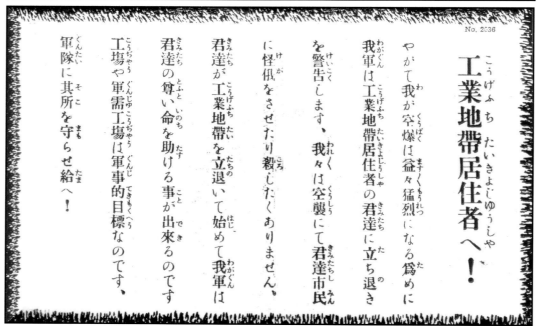

Reverse Side of Leaflet

DANGER!! Factory and production cities. To the residents of industrial districts. In the course of time our air raids are to become more and more severe. Our forces are issuing a warning to your people who are residents of industrial districts. We do not wish to hurt you or to kill you civilians through the air raids. Only upon your evacuation of the industrial districts our forces will be able to save your precious lives. Factories and munition plants are our military targets. Therefore you should let your military defend such places.

Face of Leaflet

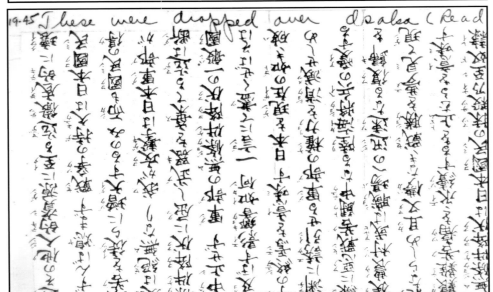

Reverse Side of Leaflet

To the various people of Japan. We are presenting you a message to you from the President of the United States, Harry S. Truman. The German Nazi regime has been destroyed. The people of Japan must recognize the enormous attacking power of the U.S. Army, Navy, and Air Force. As long as your country's statesmen and military authorities are to continue the War, our aggressions will be expanded and strengthened more and more in terms of destruction and activities. The munitions production, transport, and other resources for people shall be completely destroyed, and further continuation of the War shall bring about increased suffering.

There shall be nothing to be gained by the people of Japan. We shall absolutely not stop our attacks until the Japanese Military Authorities shall declare an unconditional surrender, and throw away their arms. The Japanese Military Authorities unconditional surrender would bring about the civilian population of Japan the end of the War, and the eradication of the power of the Military Authorities that lead the country to the present day ruins, and a swift return of the Army and Navy soldiers who have been fighting a desperate fight, to their families in farming villages or work places.

Furthermore, it will stop the empty dreams of winning the War through excessive difficulties and suffering. However, we make it very clear that we will refrain from the meaning in which the unconditional surrender will lead to the obliteration or slavery of the Japanese people.

Face of Leaflet

日本國民に告ぐ

Reverse Side of Leaflet

12 circles contain the names of 12 cities scheduled for bombing in the coming air raids. From left to right, these cities are: Tokyo, Ujiyamada, Tsu, Kooriyama, Hakodate, Nagaoka, Uwajima, Kurume, Ichinomiya, Oogaki, Nishinomiya, and Aomori. (see picture on page 8)

WARNING TO THE PEOPLE

Don't you think you want to save the lives of yourself and your parents, brothers, and friends? If you do, please read this leaflet very carefully. Within a few days, the U.S. Air Force shall bomb the military facilities of four or five cities out of the cities listed on the backside of this leaflet. There are military facilities and war material production plants in these cities. The U.S. Air Force shall destroy each and every war equipment facility used by the Japanese military authorities to prolong the war with no possibility to win. However, the bombs do not have eyes to see, and therefore, there is no telling as to exactly where they actually end up landing. As you know already, America believes in Humanitarianism and does not want to hurt innocent people. Therefore, please evacuate the cities listed on the backside of this leaflet.

You are not America's enemies at all. The Military Authorities which caused you all to be sucked into the war is the very enemy of America. America considers peace to be liberating all from the oppression imposed on you by the military authorities. If we do that, a far better Japan will be completed. How would you like to support a new leader, who is inclined to stop the war, thus restoring peace. It is possible for cities not on the back of this leaflet to be bombed, at a minimum, however, 4 cities out of the cities listed on the backside of this leaflet shall be bombed. Since we are warning you in advance, we ask you to evacuate the cities listed on the backside of this leaflet.

Author's note: Photo above described on page 8

Unconditional surrender simply means abandonment of arms. The U.S. president Mr. Harry S. Truman is sending you his warning to the people of Japan, said "the term unconditional surrender" simply means "abandonment of arms" and is a pure military expression and does not refer to total elimination of the Japanese people or to the enslavement of the Japanese people. This type of thinking was imposed upon you by the former Prime Minister, General Koiso, who purposefully twisted the true meaning, in order to force the continuation of the war which has no prospect of being won.

昭和二十年五月七日獨逸は遂に無條
件にて聯合軍の軍門に降つた。
寫眞は獨逸軍参謀長ヨドル大將が
獨軍を代表して降伏狀に署名する所。

獨逸國民は満面に微笑をたゝへながら
降伏の白旗を打振り、米第三軍第九十
四歩兵師團の將兵を迎へた。

On May 5, 1945, Germany finally surrendered unconditionally to the Allied Forces. The photos shows the German Military Field Marshal General Gustav Jodl, signing the surrender document, representing the entire German Forces. The German people welcomed the soldiers of the U.S. Third Army 94th Infantry Division by waving the white flags of surrender with beaming smiles all over their faces.

"The B-29 was the biggest & best thing the Air Force had; it was the top of the line. This thing evens the score."

— Lt. Col. Ray Clinckscales

After 55,000,000 casualties in WWII Gen. Douglas MacArthur announced "these proceedings are closed."

"If there had not been a Pearl Harbor, there would not have been a Tokyo Bay"

— Source Unknown

"The atom bomb may have saved millions of lives."

—President Harry S. Truman

Chapter 3
Cold War

• Lt. Jacques Tetrick   • Tunnel

At the outbreak of the Korean War in June 1950, this country again mobilized, including a sizable B-29 fleet. This fleet had been on active duty, or they came out of "cocoons" for restoration and modernization by plants at Marietta, GA, Davis-Monthan AFB, Tucson, AZ , and Warner Robins AFB, Macon, GA. *Sweet Eloise* came out of retirement from Warner Robins where she had been "mothballed."

Most of the renovated B-29s were sent to Yokota, Japan, or Kadena, Okinawa, for operations against North Korea. The B-29 was once again thrust into battle, and for the next several years was effectively used. The aircraft played an important role in Korea, dropping more tons of bombs on Korea than they did on Japan. Thirty-six aircraft were lost, thirty of which were shot down and the others lost in operations. The North Koreans began to employ MiG-15 jets during the second year of the war, and the B-29s were no match for them. They were gradually withdrawn from bombing activities.

Capt. Jack Tetrick crews
Low-Altitude-Human-Pick-Up-Crew Eglin AFB. FL
Pilot: Capt. Jack English, Co-pilot: Lt. Jacques Tetrick,
Navigator: Capt. Edward Wilcox
Radar: Capt. Gould Cline, Engineer: S/Sgt Edmund Hart,
Radio: A1C Harry Randall Gunner: A1C Francis Conley,
Gunner: A1C Joseph Silva, Gunner: A1C Calvin Padgett.

(Not Pictured)
Russian Survelance Crew, Molesworth RAF
Aircraft Commander: Capt. Jacques K. Tetrick (Col USAF Retired), Co-pilot: Lt. James W. Tresner,
Co-pilot: Lt. Arthur L. Vikse, Navigator: Maj. John B. Tarver (Lt Col USAF Retired), Radar Navigator: Lt Donald H. Milliken (Maj USAF Retired), Navigator/ Bombadier: Lt Robert G. Lee (Col USAF Retired), Engineer: TSgt Ralph W. Maloney (MSgt USAF Retired), Radio Operator: A1C Roger Brown, Gunner: A1C Donald H. Freer (MSgt USAF Retired), Gunner: A1C Franklin Annis.

Hoof Hearted - note horse on nose
by Jack Tetrick

Assignments:
Mountain Home AFB	Idaho	
Great Falls AFB,	Montana	
Eglin AFB,	Florida	Low Altitude Human Pick up Rescue
Molesworth AFB,	England MATS	*582nd* Air Resupply,
Intelligence Service	Russian Perimeter	

Low-Altitude-Human-Pick-Up-Human-Rescue
by Jack Tetrick

Sweet Eloise was not involved in the Korean War. She was involved in an experimental project at Eglin AFB, Florida, and other smaller experimental work, starting in 1951. *Sweet Eloise* was the *only* B-29 to be modified to perform "low-altitude-human-pick-up-rescue."

These modifications were done at Brookley AFB, Alabama, and consisted primarily of installing a piece of plate glass across the nose (changing the classic B-29 nose) to give the pilot better visibility while skimming the ground. The aft bottom turret was also enlarged to permit dropping people and supplies, or to winch a person up into the aircraft. Basically, there was a "tail hook" to snag a rope stretched between two posts, which was rigged to the person. The B-29 would fly precariously low and slow, snag the rope, and winch the person up into the aircraft.

Lt. Jacques Tetrick

This experimental technique was developed by *Sweet Eloise* at Eglin AFB in Florida. Her crew, including co-pilot Lt. "Jack" Tetrick first practiced snagging dummies, then 300 lb. pigs, then a volunteer, who reported that the G -Forces were not severe. One pig got mad or frightened and got loose inside the aircraft, causing some damage. He was then given a tranquilizer injection. It was an excessive dose, and the pig died. The squadron dined on fresh pork barbecue for several days after. The volunteer pick-up-ee was Col. Howard Harris, USAF Ret., deceased.

This technique was used successfully in Korea when three C-47s picked up persons behind the enemy lines. It was also used by Lockheed C-130s to catch satellites coming from out of space over the Pacific Ocean. *Sweet Eloise* never was required to use this technique in operations. However, the Russians were aware of *Sweet Eloise* and knew that she had the capacity to drop-off or pick-up operatives behind the Iron Curtain at any time and almost any place.

It was during this era that she had the nose art of a red horse, and the name of *Hoof Hearted* painted on her nose. When this name is spoken rapidly, it becomes offensive. However, nose art was not condoned at that time and it was removed before deploying to England, and her name reverted to *44-70113*, or *0113*.

After these experiments at Eglin AFB in Florida,

she was sent to Mountain Home AFB, Idaho, in 1953 where three Wings were formed and trained. She was assigned to the 582nd ARC Wing, completed training at Great Falls AFB, Montana, which was then re-designated 582nd Air Re-Supply Group. She was transferred to Molesworth RAF Station, England, early in 1954, in the 582nd ARS Squadron. She was very active in the Cold War. She was under the command of Capt. John English. Her co-pilot was 1st Lt. Jacques Tetrick, who became aircraft commander upon English's retirement. While based in Europe, her primary duties were to wage psychological warfare against the Communists and to help counter that Communist aggression which was threatening to engulf the world.

The Russians had shown their hostility toward the "West" at the Potsdam Conference in 1945, then during the partitioning of Germany and Berlin, then their declaration of war against Japan in the final few weeks of the war. Later they caused the Berlin Airlift. They broke many treaties and caused some uprisings in Europe, and seemed bent on world domination. They had several successful "coup d'etats." They incited the Cuban Crisis and caused the Bay of Pigs. There was real fear of atomic, rocket, bacterial, nerve gas, terrorist, and all kinds of warfare perpetrated by the Communists behind their Iron Curtain. Our surveillance over-flights by the U-2 aircraft were in response to Russia's hostility and aggressiveness. This became known as the "Cold War."

Sweet Eloise was capable of guerilla operations, psychological warfare, clandestine operations, propaganda warfare with leaflet drops, border surveillance, and penetration of unauthorized air space if needed to test Soviet reactions. She dropped supplies along the Iron Curtain to friendly personnel in Czechoslovakia, Hungary, and East Germany. She conducted numerous exercises with NATO, and monitored Russian communications. The Squadron worked closely with the Green Berets and the CIA. Of course the Russians were well aware of the capabilities of this unusual aircraft.

She ended her long career in late 1956 and was decommissioned at Aberdeen Proving Ground, Maryland. Lt. Jack Tetrick had flown *44-70113* for the unusual span of over five years. He stayed in the Air Force and retired as a Colonel. He was one of Alton Evans' volunteers who moved her from

Aberdeen to Florence, SC. He presently resides in Daytona Beach, Florida, and has been to Marietta to visit his old plane.

Sweet Eloise played a distinguished role in two wars. She was among the last of the B-29s to retire. The B-29s were now obsolete after about 15 critical years. They were replaced by the B-47, a Boeing six engine turbo-jet. These were built by Lockheed in Marietta, and by Boeing at Wichita.

Old *44-70113*, or *0113*, or *Z-Square 58*, or *Z-58*, or *Marilyn Gay*, or *Hoof Hearted*, or *Sweet Eloise* had finally come to the end of her road. At Aberdeen she was hauled to a grass field and left to the elements, wild animals and vandals, along with more than 500 B-29s and other aircraft. Most of them were scheduled for scrapping or to be used as the targets for new weapon-systems. Most of the old warbirds were available for static display or for purchase at a very reasonable price to anyone who would give them a permanent home. Here she was to remain for 15 years until Mr. Alton Evans read about "his" old airplane in an Air Classics magazine in December 1972.

44-70113 *Sweet Eloise* **at Aberdeen Proving Grounds, MD.**
photos by Alton Evans

The Unique Bomb Load Feature and Tunnel of the B-29 Superfortress

Cross section through the Bomb Bay

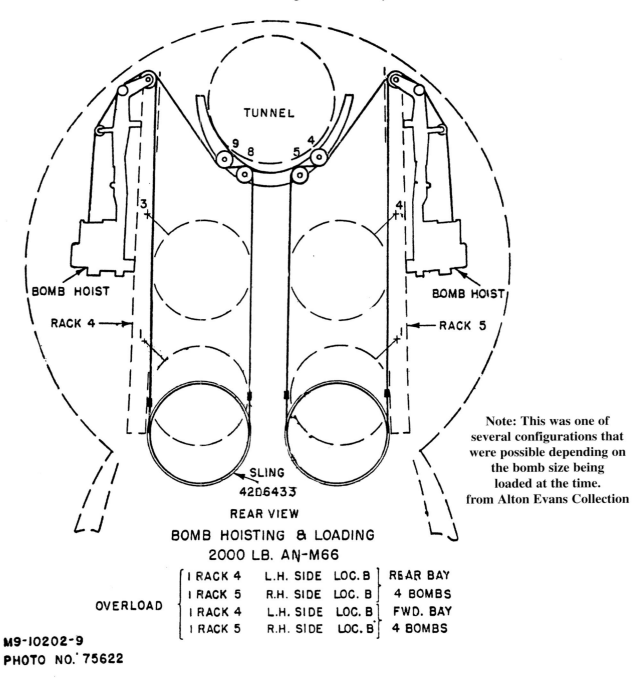

TUNNEL

9 8 5 4

3

4

BOMB HOIST

RACK 4 →

BOMB HOIST

← RACK 5

Note: This was one of several configurations that were possible depending on the bomb size being loaded at the time.
from Alton Evans Collection

SLING
42D6433

REAR VIEW
BOMB HOISTING & LOADING
2000 LB. AN-M66

I RACK 4	L.H. SIDE	LOC. B	REAR BAY
I RACK 5	R.H. SIDE	LOC. B	4 BOMBS
I RACK 4	L.H. SIDE	LOC. B	FWD. BAY
I RACK 5	R.H. SIDE	LOC. B	4 BOMBS

OVERLOAD

M9-10202-9
PHOTO NO. 75622

The unique tunnel in the B-29 permitted crewmen to crawl between fore and aft crew compartments obviating oxygen mask and heated suits. Thus the aircraft could operate at extreme altitudes with heavy loads while allowing the crew comfort and efficiency. The tunnel was 34 inches in diameter and 33 feet in length. (See pages 7 and 8 with bomb bay doors open, tunnel entrance in cockpit page 51 and tunnel entrance in aft fuselage page 52).

Chapter 4 —————
Aberdeen to Florence ————

• Alton O.Evans

Alton O. Evans is the author of a sister book, "B-29 44-70113" which is about *Sweet Eloise*. See bibliography.

Alton is an aviation consultant and owns Aviation Services. He is one of the founders of Kiwi Airlines. He is a former member of the U.S. Air Force, and employee of Eastern Airlines. He is "Mr. B-29" and has visited each existing B-29 in this country and has a file on each one. Al Evans was a ground crewman on *Sweet Eloise* during her Cold War activities. Like most air crewmen, Al developed a love and attachment with "his" aircraft. Al had an intimate relationship with her for about four years. In December 1972 he read an article in

Air Classics magazine about many aircraft being stored, or being sold, or scrapped, at Aberdeen Proving Ground in Maryland. The article mentioned 44-*70113,* which was his old airplane, and how she was scheduled for being scrapped unless a permanent home could be found for her.

Al flew into action. He was determined that he was going to save his airplane at all costs. It became an obsession with him that has lasted into the present. If it had not been for the determination of Alton O. Evans, *Sweet Eloise* would now be SCRAP!

Al contacted Mr. Tommy Griffin, a former B-29 crewman stationed on Saipan. He was the curator of the Florence Air and Space Museum in

STOP PRESS. Just before going to press we received word from Air Guard Capt. Jim Turner that is of importance to all Warbird enthusiasts. Jim reports that at Aberdeen Proving Grounds, Aberdeen, Maryland, is a unique collection of rare aircraft that will be scrapped if new homes for them cannot be found. Included in this group are eight B-29s, 44-61975, 44-70113, 44-84053, 44-61671, 44-61739, 44-27343, 44-87627, and 44-62203. The B-29s are in fair to good condition and have been sitting in a field for a number of years. The engines have very few hours on them. The Florence Air And Missile Museum has claimed several of the B-29s for preservation. Also available are two rare Martin AM-1 Maulers, one in poor condition and the other in good shape. Also at the base is the Douglas XBTD-1 but this has been claimed by the Florence group. There were over 500 B-29s at Aberdeen but they have been gradually scrapped over the years with 7 fatalities among the scrappers! Due to this fact the Army would like to give away the remaining Warbids to qualified individuals or groups. Even if you can't obtain one of these machines why not drop a line to Aberdeen expressing concern over the scrapping of historic aircraft? ●

FLORENCE'S B-29—Jim Turner of the Florence Air and Missile Museum, is keeping us informed on the saga of their B-29 44-70113. More records have come to light revealing that the plane served with the 500th BG, 883rd BS, coded Z-58 with the name "Ancient Mariner." The B-29 completed 26 combat missions and six POW supply missions and is credited with three victories. The plane has now been moved to a point at Aberdeen where it can be taken apart for the journey to Florence, South Carolina. Jim has also found a photographer who took several hundred photos, including color, of 44-70113 during its combat career. Jim reports that our mention of surviving historic aircraft at Aberdeen caused a deluge of letters. The military told Jim that letters should be channeled through the following address: Aberdeen Historic Aircraft, Attn. AMSAV-QMA, Commander AVSCOM, P.O. Box 209, St. Louis, Mo. 63166. There are now only three B-29s left, 44-62220, 44-87627, 44-61671, and Aberdeen officials would like to see all of them preserved.

WHERE IS 44-70113's CREW? Jim Turner, of the Florence Air and Missile Museum, would like to hear from the previous crew members of the B-29 44-70113 that has been obtained for their museum This particular B-29 was obtained from the Aberdeen Proving Grounds and was originally manufactured by Boeing in Wichita with Project No. 98249-R. The plane was accepted into active service on 5/4/45 at Kearney Field, Nebraska. The B-29 was then flown to San Francisco and then on to Saipan. The crew on this flight was: 1st Pilot, Gordon A. Burgess; 2nd Pilot, Jerry Hanjran; Navigator, Lewbell Bergman; Bombing Officer, James Hodde; Radar Operator, Stewart E. Jessup; Engineer, Elmer R. Phillips; Radio Operator, Ray P. Smith; Central Fire Controller, Byron Dalkert; Left Gunner, Wilbur Benjamin; Right Gunner, George Danlavich; Tail Gunner, Clinton L. Schreffer. The museum has been in touch with the Radar Operator, Stewart Jessup, and is anxious to find the rest. 44-70113 was phased out of service on 22 June 1958 at Molesworth, England, making it one of the last active B-29s. If you know of any of the crew members please get in touch with this column.

Ominous news about 44-70113 from Jim Turner
B-29 44-70113 by Alton Evans

Florence, S.C. Tommy was trying to find a B-29 for his museum. Al was able to get permission from the U.S. Air Force Museum in Wright-Patterson, Ohio, to move old *44-70113* to Florence, S.C. if it didn't cost the taxpayers anything.

Sweet Eloise was in surprisingly good condition after fifteen years of being exposed to the weather and wildlife. She had become the home for birds, bees, and raccoons. Her engines had relatively few hours on them. Al Evans thought she might conceivably be flown from Aberdeen to Florence, but regulations would not permit this.

Al undertook the formidable project of disassembling, then hauling, then reassembling her, all with only contributions, donations and volunteer labor. Al knew that it would require an extended period of time. In addition to labor he required trucks, cranes, front-end loaders, jacks, stands, tools, machinery, and money. He did not have much help from governmental agencies or big business or national organizations.

In 1973 Al was able to secure the assistance of 25 volunteers from fellow-employees at Eastern Airlines and former Air Force buddies from the Cold War. With 732 man hours of work and with eight large trucks, they were able to move most of the component parts to Florence S.C.

One of Al's volunteers was Col. Jack Tetrick, from Daytona Beach. Jack was the co-pilot, then the Aircraft Commander, of *Sweet Eloise* at the time that her name was *Hoof Hearted*.

In 1979 Al and his volunteers went back to Florence to complete the move. They had been unable to locate a truck long enough and strong

enough to haul the center section, which was 83 feet long. He finally located a telescoping flatbed truck, which was donated.

SWEET ELOISE At Florence, S.C.
by Jack Tetrick

SWEET ELOISE At Florence, S.C.
by Wanless Goodson

SWEET ELOISE At Florence, S.C.
by Jack Tetrick

SWEET ELOISE At Florence, S.C.

SWEET ELOISE **At Florence, S.C.**

SWEET ELOISE **At Florence, S.C.**

Now all the component parts had been transported to Florence, but they were unable to reassemble the plane because the Florence Air Museum wanted to move to a new location closer to the new Interstate 95. However, after a delay, the museum was unable to relocate.

Finally, after the component parts had been sitting around for all these years, Al, with a third group of volunteers, was able to start the difficult job of re-assembly in 1989, 16 years after he had first started the move from Aberdeen, and 10 years after he had completed the move to Florence. The 73rd Bomb Wing contributed over $1,000 toward the restoration.

Things were progressing nicely until disaster occurred. Not enough large bolts had been replaced in a weakened portion of the fuselage. When she was being towed to her display position, she suddenly collapsed. She broke into three pieces in the middle. This drove the propeller blades three to five feet into the ground. Al, who was inside, was not hurt and neither were several other workers. They did report a tremendous roar as she collapsed. Plans for restoration were stopped after this heart-breaking set -

back. After 16 years, Al finally gave up his dream, which is described in his book, B-29 44-70113.

About one year later, Hurricane Hugo tore through the area and did further damage. Her tail section was turned over, and the left side of her fuselage was damaged.

According to my wife, the aircraft looked like a "pile of junk" and was about ready for the bulldozers and a B-29 graveyard. *0113* appeared to have crash-landed. Al Evans and his volunteers had done a tremendous job. They had spent a lot of money and a lot of hours, and a lot of dreams and now it appeared to be in vain. They had done their best to preserve a piece of history for future generations. The aircraft looked like a derelict that nobody wanted, except for the four valuable engines and the parts that could be scavenged.

Current generations are largely not aware of B-29s and how much this country owes to it. If it had not been for Mr. Alton O. Evans and his volunteers, "his" airplane would have been scrapped in 1973, and there would be no *Sweet Eloise*. Now, *Sweet Eloise* needed another rescuer, and he appeared in 1993. His name is Mr. Coy Short from Marietta, Georgia.

SWEET ELOISE **At Florence, S.C.**

Pete Inglis in Aft Section, Macon Tech 1995

"This is the largest non-flying restoration job in history."
— Chief Louis Boos, Dobbins ARB

Aft Section, Dobbins ARB 1997

Chapter 5
"Marietta Needs a B-29

• Bill Kinney

Editorial, by Bill Kinney, published in the <u>Marietta Daily Journal</u>, July *5,* 1992

It's been nearly forgotten around these parts, but it's a fact: Mariettans built the weapon that delivered the weapon that ended World War II.

That's right. The atom bombs dropped on Japan were delivered to Japan in the bellies of B-29 bombers, *665* of which were built in the Bell Bomber Plant in Marietta.

Now, with the 51st anniversary of the war upon us, it is only fitting that one of these mighty warplanes be returned to Marietta and put on permanent display.

A small group of local airplane aficionados have been saying the same thing for the past eight years or so. The ad hoc group has had no luck, so far, at finding an available B-29, but has not given up hope.

Of the B-29s built at the Marietta plant during the war, and of the 3,960 built nationwide, fewer than 40 are known to survive. Most are on display in various museums around the world. Only three are housed in an indoor setting with the rest - all of them nearly 50 years old - exposed to the elements.

There is one B-29 still flying. She is named "FiFi" and is a part of the Confederate Air Force based in Midlands, Texas.

Sadly, thousands of B-29s that had been "mothballed" by the Air Force were scrapped about 10 years ago. If only we had known.

The B-29 was the biggest, fastest bomber with the longest range of any built by any country during the World War II. Winning the war in the Pacific would have been tough without them.

Marietta landed the Bell plant through the help of then Col. Lucius Clay (later as four-star general and military governor of Germany, as well as a Marietta native and grandfather of state Sen. Chuck Clay), R-Marietta, Jimmie Carmichael and others. They also were instrumental in getting the Army to build an airfield in Marietta, christened Rickenbacher Field, which has evolved into Dobbins Air Force Base.

The Bell plant was the economic engine that dragged Cobb County out of the Depression and started her down the road toward the metro colossus she is today. Without it, Cobb would likely be another Polk or Paulding county.

The plant was taken over by Lockheed during the Korean War, and we all know the rest of the story. But too many residents of what used to be

Bill Kinney, speaker   David Hartin, president

called "The Bomber City" don't know about its past, or even what a B-29 is.

That's why we need so badly to have one of those bombers on permanent display here. School children and others are already getting first-hand knowledge of steam engines from the newly restored "Aurora," and will be enjoying the old steam locomotive just restored and donated to the city by the Glover family that is mounted next to the Marietta Welcome Center. A vintage bomber would be equally exciting and educational for them.

Surely, land could be found either at Lockheed, Dobbins or next to them on which the plane could be displayed. There is also a possibility, according to Dr. E. P. Inglis of the search group, that the project could be combined with a 20th Air Force museum. The 20th was born at Dobbins, was the primary user of the B-29, and does not have its own museum.

Helping Dr. Inglis are Coy Short, Joe Daniell, Maj. Gen. Dale Baumler, Bob Mabry, Bob Bailey, Tony Serkadakis, Sen. Chuck Clay and others. They are seeking suggestions and advice on where to place such a museum and how it might be financed. For more information on how to be a part of the effort, call *436-3350, or* 428-8510.

Our community needs to set a goal of having a B-29 on display here. To do any less would be unworthy of those who earned for Marietta the sobriquet "The Bomber City."

Chapter 6

B-29 Superfortress Association Steps In

• Eloise Strom   • Kee Bird

Mr. Coy Short is the first and only president of the B-29 Superfortress Association, Inc. This is an organization formed in 1990 for the purpose of obtaining a B-29 for the City of Marietta. It is an entirely volunteer, community-oriented group of people who love aviation and history. They believe in keeping history alive, and they realized that this country might forget its past without continuous education. They realize that the B-29 is an important part of that history and that an intact B-29 is very hard to find. Many people, even people in Marietta, Georgia, do not know what a B-29 is, and do not know how the Lockheed plant got here, or why that unusually long stairway is on Fairground Street. The author even found a Lockheed retiree who did not know that B-29s were built here.

As the historian of the B-29 Superfortress Association Inc. I have written this book, with the Associations's approval.

Portion of B-29 Superfortress Association 1996
Pete Inglis - Bob Bailey - Bill Price - Coy Short
Bob Mabry - Joe Daniell

In Tom Brokaw's best-selling books, The Greatest Generation and The Greatest Generation Speaks he stresses that thousands of people did something special in their lives that was extremely important to our country in a very turbulent era, and they would like people to know that they had sacrificed or done something that made a difference. They would like some recognition or tribute or remembrance for this. They are proud of what they did and what they accomplished. They say we need to preserve our history. That

generation saved the world from the Axis (Germany, Italy, and Japan) when it looked like the Axis was going to defeat the Allies (Free World). The outlook was especially bleak in 1942 and 1943. The expressions "patriotism," "doing my bit for the war," "for the good of the country" "remember Pearl Harbor" were common expressions back in those days.

The B-29 Superfortress Association, Inc. deserves credit for it's contribution in this decade. The Association was able to gather a "pile of junk" together, and restore an important airplane. No other group in this country had the vision and desire to take that pile of junk and make something out of it. This airplane's rebirth has become a war memorial and a history lesson.

Eloise Strom

Coy Short is one of those persons who wants to preserve history, and to keep it alive. In addition, he wanted to recognize his mother, whom he dearly loves, and who worked at Bell where she helped build those beautiful birds. His mother, Mrs. Eloise Fisher Barfoot Short Strom, was the head of the telegraphy department at Bell Aircraft Corp. She was not a "Rosie the Riveter." At present she is working for Rich's Department Store. She is still active, attractive, and sharp. She is the largest

Telegraphy Department at Bell
Eloise Strom in white, top row
from Coy Short

benefactor. She has another son, Duke Short, who is an Aide to Sen. Strom Thurmond. They were instrumental in obtaining the transfer of *Sweet Eloise* from South Carolina to Georgia.

There was a popular wartime tune called "Sweet Eloise," which was recorded by Russ Morgan. The lyrics were written by Mack David and the music by Russ Morgan, and recorded by his Music in the Morgan Manner "Big Band" Orchestra. Russ Morgan perfected the muted - trombone "wah-wah" style and was very popular during the 40's and 50's. It is recorded on a CD by Hindsight Records Inc.

"Sweet Eloise sing the birds in the trees,
When she is near you can hear them
singing sweet melodies,
They're just for my Eloise.

Sweet Eloise is a beautiful sight.
Old mister Moon comes around to look
at her every night.
Her smile's a warm summer breeze,
The smile of Eloise.

And though there may be clouds in the skies,
There's always sunshine deep in her eyes.
In case you didn't know,
roses grow, hoping someday,
They'll be pressed and caressed, in her bouquet.

Sweet Eloise is so lovely to love,
You will agree she's the only girl
that you're dreaming of,
But you'll be wasting your time,
Cause Eloise is all mine."

Courtesy of Hal Leonard Corp.

**Eloise Short Strom
from Coy Short**

**Eloise Short Strom
from Coy Short**

In 1987 a small group composed of Joe Daniell, Lee Rogers, Bob Bailey, Tony Serkadakis, Pete Inglis, and Chuck Clay got together to discuss getting a B-29 for Marietta. They found Coy Short who agreed to be the leader. They formed the B-29 Superfortress Association, Inc. Coy is an executive at Social Security. He is on loan to the Atlanta Chamber of Commerce. He is chairman of the Military Affairs Committee, where he has contact with many military, political, business, and social leaders. Chuck Clay is the attorney. He was a state senator, and is the son of a 4-star general, and the grandson of 4-star General Lucius Clay. Officers on February 2000 were: Coy Short, President; Joe Daniell, Vice President; Barbara Shaw, Secretary; Bill Price, Treasurer; Chuck Clay, Attorney; Pete Inglis, Historian.

Other active members are or have been; Davis Adcox, Fred Aiken, H. B. Armitage, Max Bacon, Bob Bailey, Mickey Bailey, Paul Baker, Thomas Barnes, Stacy Bass, Gen. Dale Baumler, Frank Beadle, George Beggs, Maj. Michael Bistrica, Charlie Bollech, Louis Boos, Clyde Bramlett, Bruce Bromell, David Brown, Charlie Brown, Johnny Browning, Howard Burgett, Bill Byrne, Jay Cahill, Jeff Cain, Bill Calloway, Bob Christian, Beau Clark, Robert Clifford, Victor Colborn, Tim Collins, Jesse Colvin, Mike Connelly, Paul Connelly, Dan Cox, Sen. Paul Coverdell, Dr. Harlan Crimm, Herman Daniell, Rep. Buddy Darden, Bill Dean, Jr., Richard Dennison, Dobbins Thrift Shop, William Dunbar, Bill Dunaway, Eric Duvall, Col. R. A. Ebert, Howard Ector, William Edwards, Jane Eisele, Calvin Ellington, Bob English, Robert Eskew, 504th B.G. 58th Bomb Wing Assn., Thomas Faustino, B.J.Fleming, Buck Ford, Ray Fowler, Bill Franklin, Gene Frazier, Michael Fuchcar, Robert Gaines, Bob Garrison, Rep. Newt Gingrich, Art Gizynski, James Bolan Glover, Dr. Gordon Gurson, Francis Gore, William Greenwood, Bill Gustafson, Hap Halloran, Elmo Harrell, Dr. George Harrison, Dr. Bob Hays, Mark Herring, Charles Hill, Marvin Horner, Louis Hohenstein, Jack Houston, George Hyland, Harry Ingram, Ira Iburg, James Janeway Jr., John Jemmy, Sr., Gilbert Johnson, Robert F. Jones, James H. Kelly, Joe Kelly, Walter Kelly, Monroe King, Bill Kinney, Joe Kirby, Dempsey and Corene Kirk, Mrs. C.H. Kohler,Jr., Joseph Konchalski, Dee Langley, Marsha Lanigan, Robert and Edith Latsch, Carl Leighty, Pat Leverentz, Carl Ligman, Harry Livingston Jr., George Lucas, Bob Mabry, Tom Major, Bill Maloney,

Charles and Dorothy Maloney, Levin Manly Jr., Jim and Ellen Marler, C.W. Marlow, Sherman Martin, Glynn Mathis, Steve May, Lloyd McCall, S.S. McDonald, Jim McDuffie, Charles McIntire, Bob McPherson, John Matlock, Mayor Ansley Meaders, Jack Millar, Sen. Zell Miller, Travis Mobley, Douglas Moore, Jim Moore, Dick Morawetz, Ed Mulkey, John Nino, Sen. Sam Nunn, Merrill Nuss, Guy Northcutt, Jr., Officers Wives Club, Dick O'Hara, Martin O'Toole, Becky and Bill Paden, David Paggi, Robert Peck, Capt. R.D. Peterson, Virginia Petree, Don Poor, John Powers, John Pruett, Pete Reeve, C.V. Reeves, Wayne Reeves, Judy Renfroe, Jack and Barbara Renshaw, Eugene Rhodes, Dickie Roberts, G.C. Robinson, Paul Robinson, Dick Rogers, Lee Rogers, W.C. Russell, Jr., Paull Saffold, Steve Schmidt, Dr.Tom and Kathy Scott, Lee Secrest, Dr. Phil Secrist, Emil Seiz, Tony ànd Faye Serkadakis, Harold Shamblin, Robert Shear, Bob Shearer, Cliff Shirley, Richard Sims, Roy Simmons, Jr., David Skillng, Fred Smith, Harold Smith, Lucia Smith, C.T Smithweck, Bob Snapp, Darrel Steams, Robert Stephens, Louis Sohn, Virgil Spence, Skipp Spann, Rudy Stankus, Mac Stamper, Regina Goldsworthy and Norman Stott, Joanne P. Stratton, John Taylor, Daniel and Ada Thompson, Gen. Seymour Thompson, Hurth Tompkins, Steve and Virginia Tumlin, Gary Vollmer, Oliver Welch, Ernest Wester, William White, Mrs. Jerry Whiteman, Mary Wiley, Dr. Sid Williams, Maggie Willis, Richard C. Wing, Howard Wolf Jr., Dianne Woods, Wade Woodward, Jim Wride, David Young, Dr. Koji Yoda, Dr. Henry Zimmerman, and many others.

Several abortive attempts were made to obtain a B-29. After the war there was a tremendous surplus of aircraft, but we started our search decades too late. We first made inquiries in regard to the *Enola Gay*. We next inquired at Davis-Monthan AFB and Pyote, Texas. We tried to obtain *Dark Slide*, the one west of Cordele at the Georgia Veterans Memorial State Park. We also inquired from Col. Kermit Weeks at Weeks Air Museum in Miami, and then China Lake NAS, California. Having no success, we then made another attempt to obtain the one near Cordele.

Warner Robbins AFB GA
Marietta Built Model B
Moved into hanger

Cordele B-29
F-13 version at
Georgia Veterans Memorial State Park

Kee Bird

Then we learned of *Kee Bird* which had crash-landed on a shallow frozen lake in northern Greenland in 1947. We determined to obtain it. Sen. Sam Nunn and Rep. Buddy Darden were most helpful. We obtained permission from the Danish government, which owns Greenland, and from the US Air Force Museum at Wright-Patterson, Ohio. We were going to use a helicopter, a barge, and a C-5A to bring it from Thule AFB in 1991. Unfortunately the Persian Gulf War thwarted our plans.

A group from Southern California, headed by Darryl Greenameyer, performed a prodigious restoration job on *Kee Bird* in Greenland in 1994-95. While doing a high-speed taxi run on the frozen lake in which she was nearly air-borne, the APU (auxiliary power unit, the putt-putt) bounced loose due to rough ice, and caught fire. The entire plane was consumed by fire. The story of *Kee Bird* is a heart-breaker. It is on the video-cassette "Frozen in Time," and may be purchased from PBS, Nova.

Our Association learned from the Air Force Museum that a B-29 might be available in Florence, S.C.

Kee Bird burning on Greenland lake
by Tim Wright in <u>Smithsonian</u>, September 1995
©1995 Tim Wright/www.timwrightphoto.com

Chapter 7
Restoration

- Georgia Dept. of Vocational and Adult Education
- Dobbins ARB
- Warner Robbins AFB
- Nose Art
- F.A.Q. Web-Site

In 1993, the small B-29 Superfortress Association, Inc. learned that a derelict B-29 was available, located in an air museum in Florence Air And Space Museum, S.C. Our Association believed we could move, restore, and re-assemble this aircraft for display. She was and still is owned by the U.S. Air Force Museum in Wright-Patterson, Ohio.

President Coy Short and the Association presented our proposal for their consideration. When they learned that our assets were only $35, they didn't take us seriously. We started looking for benefactors, and came up with $10,000. They changed their tune, and we soon had one big, old "pile of junk" B-29 under our wing, to love, to fix up and to take care of, for the rest of her life. We convinced them that we had the capacity, the determination, and the enthusiasm to recover it, to restore it for public display, and to complete the project before the 1996 Atlanta Olympic Games.

Georgia Dept. of Technical and Adult Education

Dr. Harlan Crimm came up with and presented to Joe Daniell the concept that this should be a state-wide project, and that we should enlist the help of the state of Georgia. This would benefit the entire state, and hundreds of Georgians would become involved with the whole project. This was a fantastic concept. Dr. Crimm contacted Mr. Bob Mabry, Deputy Commissioner of the Georgia Department of Technical and Adult Education, or the "tech schools." He felt that the aviation technology programs in the technical schools could help with the restoration.

There are thirty-five vocational schools scattered throughout the state of Georgia. Seven of these offer programs in Aircraft Structural Technology, three in Aviation Technology and two in Avionics. These schools need aircraft on which to work and to learn. One of the courses includes the restoration and repair of vintage aircraft. Another department is truck driver education. There are nine truck-driving programs, and they too need unusual loads on which to learn and

practice. Joe Daniell and Dr. Crimm met with Bob Mabry. Mabry received the "go ahead" to proceed with the restoration from Dr. Kenneth Breeden, Commissioner. Bob Mabry assigned Mr. Dea Pounders, president of South Georgia Tech located in Americus, to be the lead school.

Joe Daniell expanded the plan to involve an Air Reserve logistic unit from Warner Robins AFB. This unit also needed aircraft on which to learn and to keep proficiency. Then a maintenance unit from the 94th Airlift Wing at Dobbins ARB was assigned to do the cosmetic work. This consisted primarily of site preparation, landscaping, lighting, considerable sheet metal work, painting, applying decals, bird and water proofing, sealing the engines, installing the turrets and the Plexiglas blisters, and restoring the nose. Dobbins ARB was under the strict supervision of the U.S. Air Force Museum, which wanted the aircraft to be presented in the best possible light. In addition to all this aid from the military and the schools, Gov. Zell Miller and the Georgia Legislature contributed $30,000 to the project.

Bob Mabry and Mr. Pounders called a meeting of all seven presidents of the seven A and P (Airframe and Propulsion) schools, and presidents of the five trucking schools to discuss plans and to make assignments. They were given the time constraints of one and one-half years, in order to complete the project by the start of the Atlanta Olympics. Structurally, she was completed by the Olympics, but she needed much cosmetic work.

Their assignments were:

South Georgia Tech at Americus............wings and landing gear
Heart of Georgia Tech at Eastman......nose section
Flint River Tech at Roberta ...wing tips and ailerons
Middle Georgia Tech at Warner Robins ....center section and tail section
Macon Tech at Macon..........................aft fuselage
Clayton College and State U. at Morrow ....two engines and props
Atlanta Tech at Atlanta ....two engines and props

The trucking schools assignments were:

Swainsboro Tech ....nose section

Other schools involved were:

The seven schools of Georgia Adult Vocational and Education Department and the five trucking schools were involved for about one and one half years from 1995-96. Much of the work was difficult and challenging, partly due to the age and the condition of the aircraft.

The schools did not attempt to restore the interior of the plane for several reasons: they wanted to keep the weight down because of metal fatigue; there had been too much vandalism; there was too much work to do on the exterior; there was little likelihood that people would be permitted inside; the project was already very complicated; and it would add to expenses. Several small interior parts were saved for museum purposes.

The schools' basic jobs were to repair or replace the damage from corrosion, from collapse, and from the Hurricane Hugo. They were to replace skin where necessary, replace missing parts and panels, apply primer, prepare for painting, insure the fits would be strong and easy, insure structural integrity, and prepare for transit.

Following her neglect and her collapse, and then Hurricane Hugo, we were pleased that she could stand up by herself. She is not likely to collapse for many years. The workers were all careful to not damage her further since she was so fragile. She was over fifty-one years old, and had been exposed to the elements continuously. Her engines have low hours, but structurally and mechanically, she is unflyable.

In 1994, she was dismantled, and trucks were assigned to haul component parts from Florence to the various technical schools. All organizations, administrators, instructors, and students from Georgia were excited about the project. The quality of their work, attitude and enthusiasm to help restore this old warbird were impressive, according to Bob Mabry who was the state's project supervisor. When the restoration was completed, one and one half years later, trucks hauled the parts to Dobbins ARB, a major job. The trucking students were involved with the planning, loading, tie-downs, travel to the schools, insurance, accommodations, positioning in the storage area at Dobbins, and return to the trucking schools.

Dobbins ARB

The two hauls (to and from the schools) happened without mishap or damage to the component parts. The final haul occurred in January, 1996, during one of the coldest and windiest days of the year. The parts were stored at Dobbins for three months for the winter. Reassembly began in April and was basically completed in time for the Atlanta Olympic Games which were held in July and August, 1996.

It is possible that *Sweet Eloise* is the largest, non-crash, non-flying, restoration job that has ever been attempted. It is certainly one of the largest and best. She was a derelict that has been retrieved from the graveyard. The state of Georgia "turned a sow's ear into a silk purse."

Component parts at Dobbins February, 1996

Component parts at Dobbins February, 1996

Component parts at Dobbins February, 1996

Cockpit looking aft at tunnel (top) and bomb-bay entrances; throttles and pilot seat on right.

Cockpit looking forward at "greenhouse" and trap door

Nose section at Heart of Georgia Tech at Eastman (note "nose art" by student)

**Center Fuselage at Middle Georgia Tech at
Warner Robins AFB**

Outer Wing at Flint River Tech at Roberta

Wing section at South Georgia Tech at Americus

Aft Fuselage, at Macon Tech

Engine and Propeller at Atlanta Tech at Atlanta

Engine and Propeller at Clayton College at Morrow
Bill Price   Pete Inglis   Norm Stott

Nacelles at Dobbins April 1996

Center Fuselage

Wing and Nose, with men on wing

Painting the "Z"

"Clear to Taxi" by author

Aerial view at Dobbins ARB by Civil Air Patrol

622nd CLSS from Warner Robins AFB
Chief Steve Sheffield, 2nd from right, rear

Warner-Robbins AFB

The 622nd CLSS Reserve Unit from Warner-Robins AFB did a remarkable job of assembling the aircraft in just twenty one days. There were about 75 men working on her at various times, headed by Chief Sergeant Steve Sheffield. The men were housed in the barracks at Dobbins. Thanks to the excellent restoration done by the schools, the aircraft seemed to go back together easily. They started with the center section and the landing gear, then the aft fuselage, the empennage, the nose section, the wings, and finally the engines and propellers.

Chief Sergeant Louis Boos, of the 94th Airlift Wing at Dobbins ARB was in charge of the cosmetic work after the aircraft had been assembled. The aircraft belongs to the U.S. Air Force Museum at Wright-Patterson AFB, Ohio. Dobbins ARB is charged with its appearance, maintenance, and security. Chief Boos was able to restore the aircraft to historical correctness. The guns are excellent replicas, made of black plastic tubing. The four engines will turn, and could become operable after extensive overhaul. He was able to locate some authentic gun turrets at Warner Robins AFB. The decals and markings are near-perfect. He and Harold Shamblin visited all of the vocational schools at least once, serving as liaison, and quality–control officers.

The propellers were in fine condition despite being buried three to five feet into the ground when the nose section collapsed in Florence. The gun-sight in the right blister is not authentic since it consists of a cardboard tube, a rug shampoo jar, two peanut butter jar lids, two Coke bottle caps, black paint spray, and some white-out. One of the original Plexiglas side-blisters is in storage. The installed top-blister is original but the other two were donated by Burger King. Burger King ordinarily uses them in playgrounds.. They fit nearly perfectly.

The nose wheel door was fabricated by Dobbins, as well as some panels. Dobbins did extensive cosmetic work. The birds and rain-water needed to be kept out.

The classic B-29 nose had been modified at Warner Robins AFB in 1950, for the "low altitude human pick-up rescue system" function which had a plate glass window across the front of the nose for better pilot visibility. It would have been too costly to duplicate the framework and the Plexiglas nose, so Chief Boos did a good job with just aluminum and black paint. In addition, if Plexiglas had been used, the interior of the cockpit would have needed restoration because its deplorable condition would have been visible to the public.

The paint used on the aircraft is water-based and will need repainting after about five to ten years. A shiny, acrylic paint was not used because of possible damage to the environment and possibly getting spray on passing automobiles. *Sweet Eloise* gets a good bath twice a year, whether she needs it or not.

Nose Art

The "Nose Art" was done by Don Morris who is a retired Coca Cola sign painter. The voluptuous

bathing beauty has evoked no criticism from any woman's group, because it is so historically correct. Nor has she evoked any accidents by passing male drivers, to our knowledge. During the war, she had no Nose Art, but during the "Cold War" she had a red horse and the name *Hoof Hearted.* Care must be taken when this name is

Don Morris painting Nose Art

spoken aloud. All "Girlie" pictures were ordered removed from military aircraft in May, 1945, but during the Korean War, the Cold War, and the Vietnam War "The Girls" re-appeared.

Using student and instructor labor held down the restoration costs. Part of the student curriculum is restoration of vintage aircraft. Another saving was the re-assembly by the 622nd CLSS from Warner-

Robins since they are always looking for work to maintain their proficiency in salvaging aircraft. The truck driving schools are also looking for learning material.

In addition, Westover AFB in Massachusetts provided three pylons that support and anchor the aircraft. Dobbins donated men, equipment, and supervision. There were several outside contractors used. The B-29 Superfortress Association, Inc. contributed money. The Association received a $30,000 grant from the Georgia Legislature which helped pay for various incidental expenses such as rentals, fees, insurance, materials, equipment, tools, etc. *Sweet Eloise* became a statewide project. The state of Georgia was anxious to preserve such an important tool for educational and historical purposes. In addition, she could become an excellent tourist attraction.

Sweet Eloise is located just inside the main gate at Dobbins ARB. It is not necessary to go through the security gate. There is a parking area for about 50 vehicles. Dobbins ARB is located about 15 miles north of Atlanta on Highway 41 (Cobb Parkway). It is about one mile west of Interstate 75 at exit #261 on Hwy. 280.

The 94th Airlift Wing at Dobbins prepared the site, using a large amount of fill dirt to elevate the site, graded and then grassed. The area under the aircraft is covered with crushed rock. There is a brick path extending from the parking lot and encircling the plane. Shrubbery lines the entire

Nose Art   Note: 27 Bombs (missions), 3 flags (fighters), 5 parachutes (POW drops)

pathway. There is ground lighting along the entire path and four spot-lights which burn all night to illuminate the plane. She is an impressive sight, day or night.

Chief Boos states that this is one of the most extensive restoration jobs of non-flying aircraft in the country. There are no plans to build a cover for her to protect her from the elements. Aluminum does not last forever, even if painted and cared for. It oxidizes. For the long- term, she needs the minimum of a simple umbrella for protection, similar to the B-17 *Memphis Belle* in Memphis.

In addition, she needs some other things that will make her restoration complete, such as two or more benches, brochures, a plaque for each of her principal crews and their missions, another history plaque, a map on the ground under the plane from wingtip to wingtip showing the geography and distances, highway signage, and more community involvement. *Sweet Eloise* would come to life. The City of Marietta could host a reunion of all the old crewmen for some VJ Day celebration, and a get-together of all old Bell employees.

Frequently Asked Questions

Will she fly?..No
Can I get inside?...No
Who owns her?................U.S.Air Force Museum,
　　　　　　　　　　　　　　Wright-Patterson, OH
Where and when was she built?..............Wichita, KS,
　　　　　　　　　　　　　　　　　　May, 1944
How did Marietta get her ....Read this book
When did Marietta get her ..1994, Dedicated 1997
Who takes care of her?..........................Dobbins, ARB,
　　　　　　　　　　　　　　94th Airlift Wing
Who supervises her? ....Marietta's B-29
　　　　　　　　　　Superfortress Association, Inc.
How much did she cost?.......................Unknown,
　　　　　　　　　　　　　　New $665,300
How long did restoration take? ....2 1/2 years
Were B-29's important..........................Extremely
Was *this* an important airplane? ....Unusually so

Does she have an interesting history? ....Yes
Does she need a museum?...............................Yes
Does she need a cover? ....Yes
Does she need some money?...........................Yes
Where can I send it? ....B-29 Superfortress
　　Association, Box 811, Atlanta, GA, 30301
Why did you do all this?..........Education; history;
　　improve Marietta
Where can I buy this book?...Marietta Museum of
　　History, 1 Depot St. Marietta, GA 30060;
　　or author at PO Box 3292,
　　Marietta, GA, 30061-3292, autographed.
Is this book in the Cobb County Library?........Yes
Does she have a Web Site?...............................Yes
　　Dedicated to *Sweet Eloise* and to Sally Ann
　　Wagoner's father-in-law, S/Sgt Jim
　　Reifenschneider
　　http://home.att.net/~sallyann2/b29.html

Group of B-29 Superfortress Association observing restoration of Sweet Eloise at Dobbins ARB, 1996. (L to R) Chief Louis Boos, Coy Short, Harold Shamblin, Pete Inglis.

Heard and Overheard:

A receptionist at the Marietta YWCA was asked if she knew what a B-29 is? She said, "No. Is it a vitamin?" The other receptionist next to her said, "No. Is it a rock group?" Another person added, "I think it's some kind of an airplane, isn't it?"

At a restaurant, one gentleman said, "Don't tell me y'all made B-29s here. There's sure no sign of it now. That's fantastic. I'll be damned." A young person added, "That's cool."

A lady asked, "Were B-29s made here at the Glover Machine Works?"

One day after school, the mother of a teenager asked him "What did you learn in school today?"
He replied, "We learned all about WW Eleven."

All of us could use some more education, especially about an important event such as WW II and the part that Marietta and the B-29 played in the war.

***FiFi* at DeKalb Peachtree Airport, 1996   note "greenhouse"**
See pages 73 and 74

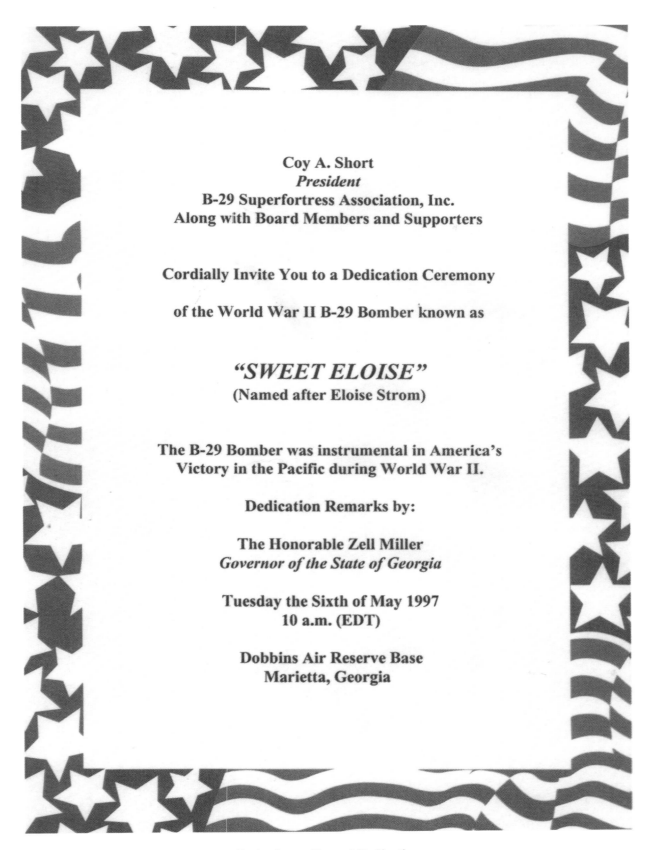

Coy A. Short
President
B-29 Superfortress Association, Inc.
Along with Board Members and Supporters

Cordially Invite You to a Dedication Ceremony

of the World War II B-29 Bomber known as

"SWEET ELOISE"
(Named after Eloise Strom)

The B-29 Bomber was instrumental in America's
Victory in the Pacific during World War II.

Dedication Remarks by:

The Honorable Zell Miller
Governor of the State of Georgia

Tuesday the Sixth of May 1997
10 a.m. (EDT)

Dobbins Air Reserve Base
Marietta, Georgia

Invitation to Formal Dedication

Chapter 8
Dedication Ceremonies

• Re-Naming Dedication • Formal Dedication: May 6, 1997

• Gov. Zell Miller Address

Sweet Eloise has had two separate dedications in recent years: one on September 22, 1994 for her name change and the start of her restoration, then another on May 6, 1997, for her formal dedication.

Re-naming Dedication

The re-naming dedication was held in a hangar at Dobbins ARB at 10 AM. The un-restored nose section had been hauled from the Heart Of Georgia Tech at Eastman. Gen. Gordon Sullivan, U.S. Army Chief of Staff, and Gov. Zell Miller were the keynote speakers. President of the B-29 Superfortress Association, Inc. Coy Short introduced his mother, Mrs. Eloise Short Strom. She was the principal benefactor of the Superfortress Association. She was also an important employee at Bell Aircraft, as the head of the telegraphy department.

Z-Square 58 (44-70113) was officially named *Sweet Eloise* in 1994. Her previous names were *Marilyn Gay, Hoof Hearted,* and again, *Marilyn Gay.*

Coy Short also introduced his brother, Duke, who is Eloise Strom's other son. Duke Short was instrumental in obtaining and transferring the aircraft from South Carolina. He is the Aide for Sen. Strom Thurmond. The transfer of the aircraft from South Carolina to Georgia was a difficult assignment.

In addition, Coy Short took the occasion to express thanks to the Georgia National Guard for their assistance in the major flood relief in South Georgia around Americus several months previously, and he

Sen. Strom Thurmond, Duke Short 1994

Re-naming Dedication, April 1994
Duke Short, Eloise Short Strom, Gen. Terry Whitnell,
Coy Short

recognized the Georgia Air Reserve and their employers. About 2,000 people attended.

The 35-foot nose section, which was the only section present, had been hauled from Eastman, Georgia where this section was being restored at the Heart of Georgia Technical School. This restoration work took about one and one half years. There were six other schools working on component parts of *Sweet Eloise* which were reassembled at Dobbins ARB before the Olympics in 1996.

Two former crewmen attended this ceremony. Lt. Wanless Goodson was the co-pilot on the Lt. Adamson crew. He pointed to an aluminum patch indicating where a bullet narrowly missed hitting his right shoulder. Wanless presently lives in Welch, West Virginia, where he owns a supermarket. His letter is printed in Chapter Two (page 28). Also attending was Cpl. Jesse Colvin who was the radar operator on the Lt. Adamson crew. He presently lives in Pulaski, Tennessee. Both of these former crewmen were very excited about what we had accomplished with their old warrior.

Formal Dedication

On May 6, 1997, the formal dedication was held at 10 AM at Dobbins ARB on a beautiful spring

day, with around 1,000 people attending. The principal speaker was the Honorable Zell Miller, Governor of Georgia, now Senator. He gave an eloquent keynote address indicating the importance of the B-29 to him and his family, to this area, and to the country. In addition, Rev. Nelson Price, of the Roswell Street Baptist Church, Col. William Kane, C.O. of Dobbins ARB, Commission Chairman Bill Byrne, John Pruett from WSB-TV, and Col. Ray Clinkscales were on the program. The "nose art" of the scantily clad pin-up girl was then unveiled by Mrs. Eloise Strom and Gov. Miller. The "nose art" by Mr. Don Morris is historically correct.

The wartime hit-tune "Sweet Eloise" was sung. Former Bell employees and former B-29 crewmen were recognized. Beautiful two-foot blown glass B-29 models were then presented to Eloise Strom, Gov. Miller, and to Coy Short. A reception was held in the Consolidated Open Mess at Dobbins ARB.

Publicity was excellent for the event. Exceptionally fine articles and photographs are available in the <u>Marietta Daily Journal</u> on May 4, 6, and 7, and the <u>Atlanta Journal</u> on May 6, 1997.

There is a small sign painted under the pilot and co-pilot's windows which is not an actual crew, but lists the B-29 Superfortress Association officers.

Mrs. Eloise Strom   Gov. Zell Miller

Gov. Zell Miller   Mrs. Eloise Strom

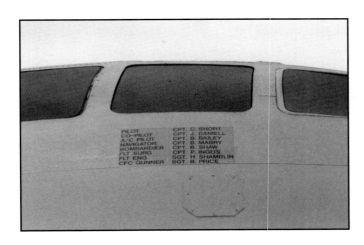

Dedication Dobbins ARB May 6, 1997

Pilot	Coy Short
Co-Pilot	Joe Daniell
Gunner	Bill Price
Bombardier	Barbara Shaw
Radio	Bob Bailey
Radar	Bob Mabry
Flight Surgeon	Pete Inglis
Mechanic	Harold Shamblin

Marietta B-29 Superfortress Association, Inc. "crew"	
Coy Short	President
Joe Daniell	Vice President
Bob Bailey	Treasurer
Chuck Clay	Attorney
Bob Mabry	Coordinator
Barbara Shaw	Secretary
Pete Inglis	Historian
Harold Shamblin	Quality Control
Bill Price	Treasurer

Sgt. Jim Reifenschneider   Capt. Pete Inglis
Col. Ray Clinkscales

Col. Ray Clinkscales was recognized. He was the pilot of *Marilyn Gay* on her first eight missions over Japan which was from May 23 to June 14, *1945.* Ray was the commander of 22 previous B-29 missions over Japan in a different B-29, *Ancient Mariner.* In addition, he had previously served a tour of duty as the pilot of a B-24 Liberator and A-20 Havoc in Africa.

S/Sgt. James Reifenschneider was also recognized. Jim was the Central Fire Control gunner on Col. Clinkscale's crew. Reifenschneider presently lives in New Philadelphia, Ohio. His daughter-in-law, Mrs. Sally Ann Wagoner, from Snellville, GA, also attended. She manages an excellent B-29 Web Site, honoring Jim and *Sweet Eloise.* The Web address is: http://home.att.net/~sallyann2/b29.html.

The former Flight Surgeon of the 883rd Squadron was at the dedication, Dr. Henry Zimmerman, who resides in Vero Beach, Florida. As the Squadron Flight

Eloise Short Strom, Coy Short, Dr. Henry Zimmerman, Audrey and Col. Ray Clinkscales and Sgt Dempsey Kirk

Surgeon he treated and looked after all of the former crewmen of *Z-58* during the war. 'Doc Z' volunteered to fly as an observer on six missions over Japan, and earned a Purple Heart for a minor injury on one mission. He was an observer on the terrible fire-bombing raid over Tokyo. "Doc Z" published a medical article about aviation battle fatigue in B-29s, based on his studies in our squadron. He has written a total on 120 medical articles. He has some excellent color slides of his flights and of life on Saipan. He retired as an eminent cardiologist, making his first sky-dive jump last year at age 83.

The former Squadron Commander of the 881st Bomb Squadron, Col. Ralph. A."Pete" Reeve was in Marietta for the Dedication. The 881st is an adjoining squadron to the 883rd. Col. Reeve resides in Roswell, GA. and he has written an enclosed letter (see page 21).

Governor Zell Miller: Dedication Address
B-29 Bomber Dedication, Dobbins ARB, May 6, 1997

"I was here in September of 1994 to join you in welcoming the nose section of this B29 bomber, which

Gov. Zell Miller

was the first part to arrive. And I'm pleased to be back today to dedicate the entire plane as it goes on display.

Later this month, we will celebrate Memorial Day, and we will remember all those who fought and died with honor to preserve and protect the freedom we Americans cherish. But today we honor a different army, another group of patriots who made a different, but very significant contribution to World War 11. Today, as we unveil and dedicate *Sweet Eloise,* we remember and honor the hard-working Georgians, many of them women, who worked at the old Bell Bomber plant, now Lockheed.

Gov Miller's address continued

The entire B-29 project cost $3 billion, and was the largest military undertaking in American history. The first bulldozer scraped the ground on this site in March of 1942, and began what was the largest construction project in the South. Some worried that this site was too close to the coast, and planes from aircraft carriers in the Atlantic could bomb it. But that fear was offset by the availability of a workforce. Many in that workforce were women, and this was the first major industry to bring women in on an equal basis with men, creating the "Rosie the Riveter" character.

It was the largest plant of its kind in the world — 73 acres under one roof At its peak it had over 28,000 employees, and in the four years it was in operation, it built 668 of these B-29 bombers. This was one of five plants around the nation that together built nearly 4,000 B-29 aircraft during World War II - aircraft that played a very important role in the Allied air attacks, both in Europe and in the Pacific. Back in 1924, General Billy Mitchell was court-martialed for saying that air power could win a war without land invasion. But the B-29 proved him right, because it enabled us to defeat Japan without ever landing troops there.

It is a special personal honor for me to be part of this celebration, because my mother was one of those women who helped build B-29 bombers here in Marietta. Of course, we saved newspapers and flattened tin cans, and reused the grease in the frying pan like everybody else. But my mother had a patriotic fervor to do more to help win the war.

I was about ten years old at the time, and my mother came down out of the North Georgia mountains with my sister and me in tow, and went to work as an inspector at the Bell Bomber Plant. She always considered her real contribution to the war effort to be her work in the B-29. Still hanging on the wall in the living room of our house in Young Harris today, is a picture of a B-29 with the name "Birdie Miller" embossed on it - a gift of appreciation from the plant to my mother when the war was over and she went back home to the mountains.

She is among those whom we remember today, together with Eloise Strom, another former Bell Bomber plant employee who put up the seed money to restore this plane. And it is named in her honor.

There are only about two dozen B-29s remaining in the world today. So it is a special privilege to have one of them here in Georgia. This particular B-29 was originally made in Wichita, Kansas. And she is a tried and true veteran, with a long military career. *Sweet Eloise* flew 27 bombing missions in the Pacific during World War II, plus a number of missions to deliver supplies to prisoners of war. Although she was hit several times by enemy fire, she never suffered serious injury, causing her crew to refer to her as "sweet."

After the war, she came home to Warner Robins for a rest, then was reconfigured and sent to England to fly spy missions in Soviet-controlled air. She was finally retired, old and weather-beaten, to Florence, South Carolina, where she suffered damage from Hurricane Hugo.

But thanks to the efforts of Georgia's B-29 Super Fortress Association, the Air Force Museum has given this B-29 to the 94th Tactical Airlift Wing here at Dobbins Air Force Base. And thanks to the efforts of the Georgia Department of Technical and Adult Education, she stands here today, refurbished and re-assembled as a tribute to those who served their country by building B-29s here in Marietta at the Bell Bomber Plant.

Tech students who were learning truck-driving skills, drove to South Carolina and picked up the plane. Georgia's Adjutant, General William Bland helped to arrange equipment to load and unload the parts of the plane at either end of the journey. The plane was divided into seven sections, and each section went to a different state technical institute. There students studying aircraft mechanics restored and refurbished the various parts. The nose cone was the first part to be completed, and over the past two and a-half years, the remaining six sections have made their way here from around the state. They were assembled by units from Warner Robins and from Dobbins into the complete aircraft we dedicate today.

So this is not a Marietta project, it is a Georgia project, the result of teamwork by people from all across the state. Just as people were recruited from all over the state to build the B-29 more than 50 years ago, so people came together from all across the state in the 90s to restore and rebuild this aircraft. Today we celebrate the result of their hard work. Together they have given us a tribute, not only to a special aircraft in our history, but also to a special time in our history when the citizens of this state came together and pulled together to make this nation and our world a better place to live. This B-29 reminds us that we need to recapture and renew that same spirit as we face the challenges of our day."

Chapter 9 ————
Museums and Displays, ————

- Bill Kinney • Kennesaw House
- Miss Jane's Country Diner

Marietta Needs a B-29 Museum

Editorial, by Bill Kinney, published in the <u>Marietta Daily Journal</u>, September 23, 1994

Without the hard work of those on the home front, this country might easily have lost World War II. Housewives, farmers and old folks trooped into factories and shipyards, where they assembled the war-making tools of what President Franklin D. Roosevelt christened the "arsenal of democracy."

One of the biggest and busiest factories was the Bell Bomber Plant in Marietta, where 668 of the giant B-29 "Superfortress" bombers were assembled. More that 30,000 people worked in the plant during the war, and their contributions were honored at a ceremony Thursday at Dobbins Air Reserve Base, where the nose section of one of the last remaining B-29s was unveiled.

The plane was rescued from the scrap heap by the men and women of the Cobb County-based "B-29 Superfortress Association, Inc." who hope to see it reassembled and put on permanent display. Among those on hand for the ceremony was Gov. Zell Miller, whose mother worked in the plant during the war.

"Of all the things she did in her remarkable life the thing she was proudest of was her work on the B-29," he told the throng of more than 1,000 who gathered in a hangar for the unveiling. That undoubtedly was equally true for the many former Bell workers who attended the event.

The plane was renamed the *"Sweet Eloise"* on Thursday in honor of Eloise Strom of Buckhead, who worked at the plant as a teletype operator during the war. Her son, Coy Short, is leader of the group that brought the plane to Cobb.

The bomber has a long and distinguished history, having been flown on 26 missions over Japan during World War II and later on spy missions in Soviet airspace during the early years of the Cold War. It was later decommissioned and left in a field at Aberdeen Proving Ground in Maryland. It was moved years later to an air museum in Florence, S.C. where it suffered further damage from Hurricane Hugo. When that museum ran into financial problems, the local Superfortress group stepped in and made possible its transfer to Dobbins.

The plane, for now, has been taken apart and its components shipped to technical-training schools around the state for restoration. When reassembled in about a year, plans are for it to be on permanent display at a proposed Cobb County museum of World War II history. Only a handful of B-29s remain, so having one on display in Cobb would be a tremendous boost for local tourism.

As yet, neither site nor funding for such a museum has been identified. The nonprofit, all volunteer B-29 group merits great praise for having found the plane and brought it to Marietta. Its restoration seems assured. Now begins the group's next challenge - helping find a home and funds for the museum it has proposed. Here's hoping its idea takes flight.

• • • • • • •

To make the *Sweet Eloise* display complete, a museum needs to be developed. A static display is interesting, but interest flags after a brief period. A static display arouses questions, but a museum provides questions, and answers, and learning, similar to this book. The more one learns about something, the more interesting it becomes, and it comes to life. One can learn about the people, the challenges, the obstacles, the triumphs, the motivation, the secrets, and the history. The display and a museum could become a destination for school classes, bus tours, tourists and veterans groups.

There is no museum in the world devoted solely to B-29s, except for a small one on Tinian Island, where *Enola Gay* and *Bocks Car* took off. The B-29 warrants a museum devoted entirely to B-29s. It has a huge page in history. It was a scientific, industrial, technological, and financial wonder. It was one of our country's mainstay weapons for fifteen years. The development and performance of the B-29 is an important story, and the story *of Sweet Eloise* is also important.

Marietta may have an ideal site for a B-29 museum. Marietta would be ideal for a B-29 as well as a C-130

airlift museum. In addition to this, Marietta could become the site of the 20th Air Force Museum. Thanks to a bond issue, Savannah, Georgia, has built a $14 million dollar 8th Air Force Museum, since that air force was born in Savannah. The 20th Air Force was born here in Marietta. According to the late Maj. Gen. Haywood (Possum) Hansell Jr. the 20th Air Force traces its origin back to the 58th Bomb Wing, which was stationed here during the early days of Bell Aircraft.

There is the possibility of having a joint B-29 and C-130 museum here someday. The B-29 Superfortress Association, Inc. has a C-130, *Ghost Rider* in its possession, which is presently located at Dobbins. The exact site for display is undetermined. This aircraft was flown into Dobbins ARB in 1997. It is unique in that it is ship "Number 10" off the Lockheed line, and has been retired. It was converted into a gunship and was very active in the Vietnam War. Since it was built at Lockheed and since the B-29 and the C-130 were both so vital to our community, it would be appropriate to put them both on display and have a common museum.

Kennesaw House

For several years the B-29 Superfortress Association Inc. had an evolving B-29 museum, but it is now gone. The "museum" started as a display at an air show in 1993 at McCollum Field outside of Marietta. Next there was a display on the Marietta Square for July 4, 1994. Then an improved display was in the Cobb County Central Library for two weeks. It was moved to Chattahoochee Tech for two weeks, to Kennesaw State University Library for two weeks, to Lockheed for two weeks, to Southern Tech for two weeks, then to Eastman, Georgia at the Heart of Georgia Tech for two weeks. It also has been displayed at two open houses at Dobbins ARB.

Each time the display was moved, it was upgraded. In each location it generated interest and followers. Brochures were picked up and questions were asked and answered. This was prior to the completion and dedication of *Sweet Eloise,* so there were many people who were anxiously awaiting her debut.

Thanks to Dan Cox, the curator of the Marietta Museum of History, the upgraded display then found a home on the second floor of the Kennesaw House in downtown Marietta. The budding museum then moved to the third floor of the Kennesaw House in 1996. By that time the display had been upgraded to a full-fledged, but small and amateurish museum.

The museum tried to tell the story of the B-29 in four parts. The first was "What is a B-29?" The second was Marietta and Bell's part in the war effort. The third was *Sweet Eloise* in combat in WW II and Cold War. The fourth was *Sweet Eloise* and her restoration and dedication.

The museum was located in one large room. We had no overhead and we had good security. Around the walls were clippings, photos, strike photos, etc. The chief attraction was centered around one of the original gunner's blisters, where we used plywood to construct a seven foot mock-up section of the fuselage with a gun site, a gunner's station, a life-size manikin gunner, a diorama of B-29s flying in formation en route to Japan, and the tunnel

Diorama at Kennesaw House. Original side blister

Open House Dobbins ARB 1996

Open House Dobbins ARB 1996
Pete Inglis   Bill Price

entrance. We were able to use many of the original instruments and pieces of the plane in our mock-up. In addition, we had three glass display cases of models, books, magazines, instruments, uniforms, memorabilia, etc.

Unfortunately for the Association, the Kennesaw House found more lucrative use for this space and we suddenly found ourselves with no home.

Miss Jane's Country Diner

Fortunately, Mrs. Jane Alexander Eisele, a longtime nurse and friend, was sympathetic to our predicament, and she welcomed us and some of our display into her "Miss Jane's Country Diner." Her restaurant was located directly in front of the entrance to Dobbins ARB, and in front of *Sweet Eloise,* the perfect location and the perfect timing. Gradually we were able to assemble a fairly credible, low cost, B-29 display inside of her restaurant.

Many of Miss Jane's long-term customers were Air Force people, and they welcomed her new decor. They found it interesting, educational, and surprising. She had some groups for lunch especially to learn about the B-29. The B-29 Superfortress Assn. conducted two of its meetings there. The situation was unique. Miss Jane loved B-29s and she loved airmen. She enjoyed being the "curator" and the tour director. She made frequent suggestions for improvement. It was all put together in an amateurish way, but it was still a good start to a real B-29 museum, or to an "883rd Bomb Squadron Mess Hall." She even served Spam, but no C Rations or rationed food. We used her pastry cabinet for a full uniform and some valuable instruments. A large sign was displayed on the outside of her diner, and we also had another sign and a six-foot plywood B-29 model which would weather-vane with each little breeze.

The "museum" seemed to have found a new home. "Miss Jane" was beginning to get some spin-off from the B-29, and we could foresee some other upgrades. Unfortunately, Miss Jane rather rapidly developed some pain in her left arm which progressed to a severe case of "shingles." Since she ran a one-woman show, she was forced to close in October, 1998, with no prospects of re-opening. Miss Jane has now leased this valuable property to a car rental. It would still make a good 883rd Bomb Squadron Mess Hall and museum.

The remnants of the Kennesaw House B-29 Museum, and Miss Jane's "883nd Bomb Squadron Mess Hall" are now scattered around Cobb County with no prospects of restoration, unless someone can revive the museum from the graveyard, as we did *Sweet Eloise.*

Miss Jane's Country Diner   note: two B-29s

**Miss Jane's Country Diner
note: B-29, Weather-vane B-29**

**Mrs. Jane Alexander Eisele - owner of Miss Jane's
Country Diner shows typical displays**

Miss Jane's Country Diner, pastry cabinet, displays

Chapter 10 —
Olympics Mural —
Bombing the Big Chicken

The First Presbyterian Church of Marietta wanted a project to enhance the City of Marietta for visitors to the Atlanta 1996 Olympic Games, a project that would demonstrate the city's interesting history. The church came up with the idea of painting a large mural on the north wall of the vacant Stephens Lumber Company building, on the major intersection of Polk Street and Church Street, two blocks north of the Square.

Mr. Paul Sherwood, church administrator, organized and supervised this big undertaking. Preston Jarvis drew the original sketches, which were then submitted to the Downtown Marietta Development Authority (DMDA). The mural sketches included the many interesting and historic places and things in Marietta, including the new F-22 *Raptor* to be built at Lockheed which will give this country air supremacy for many years to come. It is hi- tech, comparable to the B-29 fifty years previously. The DMDA approved the sketch, but disapproved the F-22.

This author felt that the mural should have a B-29 in it, since the B-29 was arguably more important to Marietta history than Gen. William Sherman, and more important than an airplane which had not even flown yet. During the war Marietta had become known nationwide as the "Bomber City" then later as the "Airlift Capital of the World", but never as the "Fighter City."

The Presbyterian's mural committee was agreeable to changes, provided I would do some of the volunteer painting on the mural, and if I would do the painting of the B-29, so I appeared at a DMDA meeting. The DMDA stated that the F-22 is not historic and was inappropriate, and that I should submit a sketch of a B-29 for the mural. My sketch was approved.

After the sketches were approved, Tom Mowery transferred the Preston Jarvis drawings onto a grid on the brick wall with the help of a computer. Dr. Jeff Comaner did much of the free-hand drawing and the skilled painting. About fifteen other Presbyterians, including the author, did the fill-in painting. This went on for about three months during the fall and winter of 1995.

Mr. Harold Shamblin, a Lockheed and Air Force retiree and an Association member, selected a photo which he blew up to about 6 feet, and traced it on cardboard, which he cut. We traced this on the wall of Stephens Lumber Company on which we had painted a nice sky-blue background. It seems that we picked the two chilliest days of the winter to do our painting. The B-29 that we painted was up high in the corner, at about the limit of our 30 foot ladder, and we were exposed to the wintry blasts.

When the mural was finally finished, many of us viewed is as a masterpiece, especially the B-29, as it seemed to roar over the Presbyterian Church. The city seemed to enjoy the mural and we assume that the Olympics visitors enjoyed it as well. Bill Kinney, in his editorial fashion, commented on the mural saying that "the B-29 was flying over the Presbyterian Church on its way to bomb the Big Chicken." (The "Big Chicken" is a 56 foot high tower vaguely resembling an animated chicken which houses a Kentucky Fried Chicken restaurant, built in 1963 and restored in 1993, located on Highway 41).

About one year later, the mural disappeared - it was painted over. The owner of the building was in the process of selling the building to the First Baptist Church. He assumed that we had painted the mural for the Olympics, and that it had served its purpose. The artists and the citizens of Marietta expressed their sorrow about the city's loss of the mural. With the mural and the museum gone, at least Marietta still has its *Sweet Eloise* and its Big Chicken.

Harold Shamblin painting B-29 on mural

B-29 over Presbyterian Church en route to bomb Big Chicken

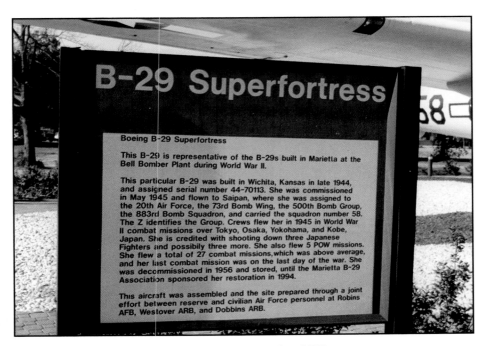

B-29 Plaque at Dobbins ARB

Chapter 11 ————————
What Is a B-29? ————————

"It has been said that the B-29 won WW II and saved countless lives because our men never had to land on Japanese soil. This war was won "in the air," and turned the tide of the entire war. We Americans owe the lifestyle we are living NOW to those who fought THEN…..an important message that has failed to be delivered to those of us who are fortunate enough to have never lived through a war. The fighting men and women and their support personnel deserve our gratitude for defeating an enemy that would have changed the face of the planet forever. We applaud those who made the ships and especially those who gave their lives so we could keep ours! It is never too late to say, Thank You." From Mrs. Sally Ann Wagoner's B-29 Website: http://home.att.net/~sallyann2/b29.html

The Need For a New Aircraft

Adolf Hitler became a mounting threat to world peace about 1933. Charles Lindbergh, the "Lone Eagle," was widely accepted by Germany and became a frequent guest of the Nazis. He became increasingly impressed with the Nazi war machine, and predicted that when Hitler was ready, he would overrun Europe and probably England. He was among those who preached "America First" and isolationism. This party felt that the U.S. should not fight in Europe's battles, but instead should build up our defenses against Hitler.

In those pre-Pentagon days, the Chiefs of Staff, Lindbergh, and others observed that our Air Force was obsolete compared to the Luftwaffe, and that we needed to build a new, large, modern Air Force rapidly. The older bombers, the B-I7s and B-24s, were excellent aircraft, but they had a distance of around 1,000 miles with a five ton bomb load. We needed a bigger, faster, heavier bomber, described as the Superbomber, or the next-generation bomber, as soon as possible.

Development of this new bomber, the B-29, was crucial. The B-29 was needed to bomb Germany from bases in Russia, or from the Azores, or from Africa. The B-36, the huge six-engine bomber still under development, was needed to bomb Germany from this hemisphere. At that time, Japan was not considered a potential enemy, even though Japan had been busy conquering most of China, Manchuria, Korea, and most of the Pacific Ocean from Alaska to Hawaii to near Australia, Indonesia and India, since 1937. Japan's empire was nearly complete.

The war started in 1939 and Adolph Hitler was conquering most of Europe, parts of Africa, Russia, Scandinavia, and the Mediterranean. During the early stages of the war, the U.S. maintained a technically neutral position, but we felt more and more threatened by the Axis, by the changing balance of world power. The U.S. became more and more supportive of the Allies, officially entering the war after Pearl Harbor in 1941. The Axis, consisting of Germany, Italy, and Japan, led by Hitler, Mussolini, and Hirohito, was now in a life and death struggle with the Allies, consisting of the U.S., England, and Russia, led by Roosevelt, Churchill, and Stalin. In 1942 and 1943, things looked bleak for the U.S. and the Free World.

The Army Air Corps realized in 1938, that it needed a very heavy, long range bomber, and in 1939, they secretly asked for bids from Boeing (builder of the B-17 Flying Fortress), Consolidated (builder of the B-24), Lockheed (builder of the P-38 Lightning, Hudson and Vega), and Douglas (builder of the A-20, DC-3 and the DC-4).

The Chiefs of Staff had the foresight to sign contracts in 1939 for the B-29, initiating the largest program of World War II, called "Operation Matterhorn," which cost three billion dollars. The second largest program was for the development of the atom bomb, called the "Manhattan Project," which cost two and one half billion dollars. Both had the utmost priority and secrecy. Both changed the world, and gave us military superiority and leadership in the world. We became a superpower.

In 1939, the Army Air Corps issued a request for design proposals for a super-bomber, "very heavy." Boeing Aircraft won the bid with a radical design for an aircraft that was almost beyond the

imagination at that time. They built a mock-up in September, 1940. Boeing received a contract to build two prototypes. They flew in September of 1942. The main contract was let in May 1941, for 250 aircraft, prior to the prototypes being completed.

For the first time in history, a major contract was made for a major high tech airplane straight off the drawing boards, before a prototype had been flown. Two hundred fifty more were ordered after Pearl Harbor. Such was the urgency that this frantic pace continued until the war ended in 1945. The race against time, which changed the world, had begun.

Design of a B-29

The fundamental problem facing the aeronautical engineers was the aerodynamic puzzle of propelling a mass more than twice that of a B-17, at a speed 50 mph faster and at a higher altitude. They needed to triple the range, and with twice the bomb load, more defensive capacity, and with greater crew comfort. This was to be the "Cadillac" of all combat aircraft. She became a very sophisticated aircraft. It was to be like a battleship, and was non-expendable. "No other aircraft ever combined so many technological advances as the B-29," according to Bombers of WW II

It was an important, complicated, and expensive challenge. There was an exasperating formula to be considered. Roughly speaking, the horsepower goes up as the cube of the velocity, which means that eight times the horsepower is needed to double the speed, instead of doubling the power. When the weight is doubled, the induced drag is greatly increased which calls for more horsepower. Is there a limit?

The answer was an unusual combination of wing configuration, flaps, power, flight and landing characteristics, which compromised neither. Enormous wing flaps, adding 21 % to the wing surface, were built, providing lift at the crucial moments of landing and take off. These were installed on a high performance wing which did not have the excellent gliding characteristics of the B-17 wing.

This monstrous B-29 was the most advanced aircraft of WW II, designed by Boeing Aircraft Corporation, intended to bomb Germany from bases in the Azores, Africa, or Russia. Boeing licensed production in four huge plants: Bell Marietta; Boeing Renton; Boeing Wichita; and Martin Omaha. With the atom bomb, it gave us world dominance.

An entirely new engine was required, far more powerful than anything previously built. The Wright Cyclone 3350, manufactured by Curtiss-Wright Corp., was developed with tremendous difficulties, which at times seemed insurmountable. This engine was called the "Achilles Heel" of the "Operation Matterhorn." There were many major delays, and at times the entire B-29 program was in jeopardy. The main problem was engine fires from heat and melting of valves, resulting from the strain of taking off, climbing to altitude, and flying at high altitudes.

Each engine displaced 3350 cubic inches, delivering 2200 horsepower on take-off. Loaded take-off and landing speeds were 120 mph, with a cruising speed of 220 mph. This engine had two dual turbo superchargers and two rows of nine cylinders. The engine approached the long-sought goal of "one horsepower per one pound."

Production of the engine finally got underway in July 1943. It was built by Chrysler's plant in Chicago which covered 6,300,000 sq. ft., and in a huge Dodge plant in Woodbridge, NJ, a suburb of New York City. These two plants produced over 35,000 engines, at a cost of $15,000 each, which does not include tremendous start-up costs. Chrysler also had a smaller plant located in Detroit. These engines were later used on the DC-7, the Constellation, the Neptune, and the C-97.

The largest propeller in the world was utilized, which was produced for the B-29 by Pratt and Whitney. It was 16 feet 7 inches with four blades. This huge new aircraft had twice the weight, twice the power, twice the load, and was much faster than a B-17, yet with the same air resistance and drag. This was accomplished by redesigning the nose, changing the wing, counter-sinking the rivets, butt-joining the sheets of aluminum, and changing the gun turrets from bubbles to blisters, which had much less wind resistance. A bubble contains the gunner and the gun, but a blister has only the gunner or only the remotely-controlled gun. The bubble could not be pressurized, but the blister with the gunner could.

The B-29 was the first bomber with

pressurization, eliminating the high altitude and extreme cold problems, such as oxygen masks and electrically heated clothing, resulting in crew comfort and efficiency. Oxygen masks and heated suits were carried for emergency use only.

The pressurization required a tunnel 34 inches high and 33 ft long, located over the bomb bays through which a man could crawl, connecting the fore and aft crew compartments. This tunnel acted like a spinal cord. The tail gunner had his own pressurization.

The 12 potent machine guns were .50 caliber and were operated by remote central fire control. Early models carried a cannon in the tail, but they were discontinued as being superfluous. Remote control means that any of the five gunners could operate any of the 12 guns, except the tail gunner. The gun sights had a primitive computer, which gave the B-29 awesome fire power. Thus these B-29s in formation were able to protect themselves quite well, and had to fear primarily anti-aircraft fire and suicidal ramming by Japanese Kamakazi fighters. They became super-powerful flying fortresses.

In the final months of the bombing campaign, as bombing effectiveness reduced the Japanese resistance, many of the B-29s were able to eliminate the cannon, reduce their armament, ammunition, armor plate, and crews, and fly at a lower altitude, and thus were able to increase the bomb loads, increase their accuracy, increase their speed and range, and decrease the strain on the engines.

The accuracy of the B-29 bombing is well known, partly as a result of the Norden bomb sight, and improved radar equipment. However, the high altitude was a problem with high winds aloft, which followed certain patterns around the world, similar to ocean currents. This was a newly-discovered phenomenon that the B-29ers named the "Jet Stream." This high altitude, sub-stratospheric jet stream interfered with the accuracy of the B-29s, and the high altitudes caused engine failures and fires.

The B-29 had many other high-tech and advanced features, and some called it a "Dream Boat." It was a super-secret aircraft. It marked the first time in history that a major aircraft was ordered into production before a prototype was flown. It was a fantastic airplane from the outset,

and the basic design was sound, but thousands of modifications were necessary. For example, the engines were plagued by engine fires. Gen. Kenneth Wolfe claimed he made over one thousand modifications in the engines alone in the China-Burma-India (CBI) theater of operations. Gen. Curtis LeMay described the B-29 as the "buggiest airplane he had ever seen."

In a sense, the entire aircraft was still being upgraded for several years after it went into production. One skeptical industrialist called the entire program "a giant mistake." Many problems were solved at the eleventh hour. Most problems could be attributed to the complexity and size, and the need to hurry into production to fight the war.

Impact of the B-29

This program would have been a major program even in peacetime. But the United States was engaged in a tremendous conflict on many distant fronts around the world, with nearly sixteen million men and women in uniform at various times. The country was mobilized for all-out war. In 1941, 1942 and parts of 1943, many people did not see any possible way in which we could win this war. We were on the defensive, and we were losing.

The United States was able to go on the offensive in 1943, with decisive victories. A planned invasion of Japan, scheduled for November 1, 1945, would have been costly and devastating to the Japanese defenders, as well as to our forces. Our Armed Forces had expanded tremendously, and were capable of mounting the invasion, but thanks to the B-29 program and the atom bomb, it was unnecessary.

Our systematic strategic bombing campaign was rapidly paralyzing, blockading, starving, pulverizing, and burning Japan. B-29s dropped 12,000 aerial mines in Japanese waterways. These mines sank more tonnage than our United States Navy did, in the final few months.

Gen. Curtis LeMay was rapidly exhausting his strategic Japanese targets and cities. Several-hundred-plane raids were becoming common, and several cities were targeted for destruction almost every night. The Japanese were suffering and starving. On March 9, 1945, 334 Superforts hit Tokyo, inflicting the greatest urban disaster, man-made or natural, in history, worse even than the

Hiroshima atom bomb. The Japanese listed 83,793 killed, 40,918 injured, 1,265,171 buildings destroyed, and 15.8 square miles of city burned in this single raid, according to <u>Trial by Fire</u>. Capt. Clinkscales and Lt. Adamson both participated in this horrific firebombing raid, but not aboard *"Sweet Eloise."* In May, 984 B-29s hit Tokyo in three nights.

However, the deadly Japanese Kamikazi suicide attacks were becoming increasingly damaging, especially during our invasion of Okinawa and Iwo Jima. During the Okinawa campaign, the Kamikazis sank 32 ships including a small aircraft carrier, and damaged 400 others. Their effectiveness was causing a major change in our strategy, which included hitting the Kamikazi bases and supplies. The Japanese lost around 4,000 of their Kamikazi aircraft, but they had 9,000 in reserve, available in the event of an invasion of their homeland.

The Japanese military was confident they could make an invasion so costly the United States would settle for less than an unconditional surrender. In addition to this huge aircraft Kamikazi fleet, they would rely on suicide submarines, mines, suicide torpedoes, suicide motor boats, beach defenses, and fortifications. They planned to concentrate on our vulnerable troop-carrier ships. However, the dropping of two atomic bombs changed their optimism.

Our Armed Forces, and our whole country, made tremendous strides toward victory after the Japanese bombing of Pearl Harbor on December 7, 1941. But the Japanese appeared almost unbeatable in 1942 and 1943. This country needed some decisive weapons to win the war. The B-29 and the atom bomb were two of these miracle weapons. Marietta played an important part of this B-29 program.

"It was estimated that a force of 1,532,000 men in all would be required for the final assault on Japan, and casualties would be heavy," according to Gen. Leslie Groves. Many veterans of WW II believe the B-29 saved their lives, because without the eight months of intensive bombing of Japan by our B-29s, followed by the dropping of the two atom bombs on Hiroshima and Nagasaki, a million or more American lives, and even more Japanese lives, could have been lost in the invasion of Japan. This amphibious invasion was scheduled to begin on November 1, 1945, on the southern island of Kyushu, and on Honshu about five months later.

Normandy and D-Day might have paled by comparison. "Japan accepted unconditional surrender while still in possession of over 2,500,000 combat-ready troops and 9,000 Kamikazi aircraft," according to Gen. "Possum" Hansell.

The U.S. Department of Defense states that were 16,353,659 people in uniform during the war: 115,185 died, and 670,846 had non-fatal injuries.

On September 2, 1945, aboard the battleship USS Missouri in Tokyo Bay, Gen. Douglas MacArthur and Japanese leaders signed the unconditional surrender. MacArthur was able to make the statement: "These proceedings are closed." Keith Wheeler in <u>The Fall Of Japan</u> described it this way: "Overhead, as if on cue, the sun came out for the first time that day, illuminating the peak of Mount Fuji and sparkling off the fuselages of 1,900 Allied planes which flew by in a massive salute. The greatest tragedy in mankind's history, one that claimed 55 million military and civilian casualties and consumed untold material wealth, and untold suffering, was finally ended. After six years, the guns were silent."

B-29 Specifications

Dimensions Wingspan ....141 feet 2 inches
Length ....99 feet
Height ....27 feet 9 inches
Weight ...Gross wt. 140,000 lb., Net wt. 72,000 lb.
Bomb Load.................................10 tons
Propellers ....16 feet 7 inches Hamilton Standard, or Curtis Electric, 4-blade, feathering and reversible
Power ....Four 2200 horsepower 18 cylinders, twin rows 3350 cu. in. dual superchargers, Curtiss-Wright
Speed........Cruising 220 mph, Maximum 375 mph
Ceiling...................................35,000 feet
Landing Gear: ....Tricycle, electric, retractable
Landing and lift-off speed ....110-120 mph *
Bomb bay doors ......Hydraulic, opens in 1 second, closes in 3 seconds
Electric motors 125
Crew 11 men. Pressurization. Temperature control
Armament.........12 .50 caliber machine guns, each shooting 700 rounds per minute. Tail cannon 20 mm. shooting 600 rounds per minute. Remote control, computer assisted. Radar. Armor plate.

Range......3,500 to 4,500 miles Operational radius, 1,880 miles

Price...$665,000

Active Duty..1943 to 1957

Number manufactured....................................3,976

Where manufactured......................Wichita 1,644,
Renton 1,119
Marietta 668
Omaha 536,
Seattle 3

* For comparison, a Boeing 747 landing/take-off speeds are 140 -170, depending on load, temperature, altitude, and wind. Limiting factor is the tires. Speeds of a Lockheed F-22 are higher, and still classified.

Results

Combat Miles FlownMore than 100,000,000 miles

Bombs tonnage ....147.000 tons

Sorties 31,387

Targets destroyed.........................40% of 66 cities
602 major war factories

Oil refinery...............83% of production destroyed

Aircraft engines ....75% of production destroyed

Cities sq. mi................175 sq. mi. 46% of Tokyo,
70% of Yokohoma

Homes .2,310,000 destroyed. 8,000,000 homeless, 21,000,000 displaced

Killed...........................900,000, 1,300,000 injured

Mines dropped.......................12,000, which sank 1,250,000 tons of shipping.

Atomic bombs...2

Planes lost overseas494 in combat,18 in operations

Planes lost in U.S...260

Losses ....averaged 1.3% of sorties airborne in combat

Aircrew lost.................3,015, dead, wounded, and missing, many of whom were executed. About 200 survived. 212 "ditched" B-29 crewmen were saved by air-sea rescue. For comparison, 8th Air Force lost 9,000 aircraft, 26,000 airmen killed, and 28,000 POWs. The casualty ratio was the worst of any branch of the service.

Korean War

Bombs tonnage ....169,000 tons

Planes Lost ....30 in combat, 6 in operations

There were 51 versions or derivations of the B-29. Important ones were: B-50;* F-13 (foto-recon); Weather; Hurricane Hunter; Search-Rescue; Tanker; SAC (Strategic Air Command); Target Tug; Mother ship for X-1; Luxury Airliner (double-decked); and Guppy (oversize freight).

*B-50 was same size; engine had 4 rows of 8 cylinders, producing 3,500 HP.

Famous B-29s: *Enola Gay, Bocks Car, Pacosan Dreamboat, Lucky Lady, Fertile Myrtle, KeeBird, FiFi, Georgia Peach, Sweet Eloise.*

FiFi

FiFi saluting *Sweet Eloise* **1996**
(Note: Just to the right of
Luke Johnson's raised right hand)

Existing B-29s

ARIZONA	Tucson	Pima Air Museum	*Sentimental Journey*	
CALIFORNIA	Atwater	Castle AFB	*Raz'n Hell*	
	Fairfield	Travis AFB	*Miss America*	
	Riverside	March AFB	*Mission Inn*	
COLORADO	Pueblo		*Peachy*	
CONNECTICUT	Windsor Locks	Bradley Air Museum		
OHIO	Dayton	Wright-Patterson	*Bocks Car*	
FLORIDA	Miami	Weeks Air Museum	*Fertile Myrtle*	**
GEORGIA	Cordele	GA Veterans St Park	*Dark Slide*	***
	Marietta	Dobbins ARB	*Sweet Eloise*	
	Macon	Warner Robins AFB	*Bonnie Lee*	****
KANSAS	Wichita		*Doc*	*
KENTUCKY	Louisville	Aircraft Industries Museum		
LOUISIANA	Barksdales AFB	U.S. Air Force Musuem		
MISSOURI	Knob Noster	Whiteman AFB	*Great Artiste*	
NEBRASKA	Omaha	Offut AFB	*Man O'War*	
NEW MEXICO	Albuquerque	Atomic Museum	*Duke of Albuquerque*	
OKLAHOMA	Tinker	Tinker AFB	*Tinker Heritage*	
SOUTH DAKOTA	Rapid City	Elsworth AFB	*Legal Eagle II*	
TEXAS	Midland	Confederate Air Force	*FiFi* only flying B-29	*****
	San Antonio	Kelly Field		
UTAH	Ogden	Hill AFB	*Haggarty's Hag*	
WASHINGTON DC	Smithsonian		*Enola Gay*	
ENGLAND	Duxford	Imperial War Museum	*Hawg Wild*	******
SOUTH KOREA	Seoul	UN War Allies Museum	*Unification*	

* *Doc* B-29 is being restored to flying condition, by Mr. Tony Mazzolini near Mojave and Inyokern, CA. sponsored by the U.S. Aviation Museum. At present she has been moved to Wichita for specialized work. Hopefully, *Doc* will be flying in 2001, and this rare bird can perform a salute to *Sweet Eloise,* as did *FiFi* in 1996.

** *Fertile Myrtle* being assembled by Col. Kermit Weeks at Weeks Air Museum.

*** *Dark Slide* F-13 version, Foto-Recon, which carried at least 25 high-powered cameras and an additional 1277 gallons of fuel.

**** *Bonnie Lee* built in Marietta. She is a stripped down B Model.

FiFi

***** *FiFi* is owned by the Confederate Air Force in Midland, Texas, a national, non-profit organization which maintains a museum and many flying Warbirds. *FiFi* has undergone major maintenance work for the past two years, and was recently touring the eastern half of this country.

She tours one half of the country on alternate years. In 1996, while en route to Atlanta's DeKalb-Peachtree Airport, we were able to divert her to make a fly-by salute over *Sweet Eloise* at Dobbins. She did a circle over Atlanta, then flew up the "four-lane" highway, dipped her wings, circled down over Marietta, saluted again, then flew to DeKalb-Peachtree. It was a very emotional moment for local B-29 lovers. *FiFi* was scheduled to fly into DeKalb-Peachtree on Oct 26-31, 2000, and was planning on performing another salute to *Sweet Eloise* and to Marietta, but while flying over Missouri she developed an engine fire. The pilot was able to extinguish the fire and to make a safe forced landing. Replacing an engine and other damage is an expensive and laborious procedure, but *FiFi* will be flying again.

****** U.S. loaned 88 B-29s to England in 1950 to aid in the Cold War. The British called them "Washingtons."

There are other B-29s which are not complete airplanes, or are at crash sites.

Chapter 12 ——————
Marietta and Its ——————
Bomber Plant

Site Selection

Because of Hitler and the war in Europe, the U.S. began to re-arm, and Roosevelt planned a massive build-up of the aviation branch of the services. The Chiefs of Staff planned to build four massive bomber plants for manufacturing B-29s. One plant needed to be in the southeast for political and economic reasons and because there was a good labor supply. Marietta became a candidate, and became the largest of the four plants.

Marietta survived the Civil War, escaping some of the wrath of the occupying troops of Gen. William Tecumseh Sherman. The churches survived the torch and served as hospitals. The Court House and all but one building on the Square were burned. Twenty antebellum houses survived. The Battle of Kennesaw Mountain was on the outskirts of Marietta. This major battle in the Atlanta Campaign extended from Chattanooga, to Chicamaugua, to Kennesaw Mountain, to Atlanta, to Savannah, and to the Sea. Marietta and Kennesaw were the jumping off sites for the famous Andrews Raiders when they attempted to destroy the railroad from Atlanta to Chattanooga.

Marietta is located on the Dixie Highway and was a main stop on the L&N Railroad. It was a one-hour ride on the streetcar into Atlanta, which was a medium- sized city of about 300,000 at the start of the war. Although the nation was in the "Great Depression," Marietta did not suffer as badly as many other areas, but unemployment was a serious problem. Many Marietta citizens objected to the bomber plant because it would bring too many changes. It would raise the cost of living and it would make Marietta a major target for German bombers and for sabotage, being so close to the coast.

Site selection for a secret Superbomber Plant was competitive. A suitable airport was crucial. Marietta had three tiny "barnstorming" airports: one off Austell Road operated by T.E.Shaw and the Hunter brothers called the Herbert Hunter Airport;

a small one west of Atlanta Road toward Smyrna, called Mosely Field; and one to the northeast of Canton Road and Sandy Plains Road intersection called McClesky Field. These were unsatisfactory.

The Army Corps of Engineers chose the present site for this major airport located south of Marietta, even though the terrain was quite rough. The Navy wanted to use the proposed airport, but the Chiefs of Staff objected to the combined use. After lengthy controversy, the Navy withdrew, and the present airport was built in approximately one year, and dedicated as Rickenbacker Field by "Captain Eddie" Rickenbacker in February, 1943. The Corps of Engineers, after a false start, built three runways, each capable of withstanding the great weight of a huge aircraft. These runways were later strengthened and lengthened to support the weight of the Lockheed C-5A transport. It was renamed Dobbins AFB in 1950 in honor of Capt. Charles Dobbins of Marietta, a fighter pilot who was shot down in the battle for Sicily.

"Marietta textile executive Guy Northcutt said, 'Whether we like it or not, our way of life in Marietta Georgia will soon be gone forever.' He predicted that old timers would be divided into those who griped about it and those who adjusted and "cashed in" on the new opportunities.' Cobb County Times editorialized that 'our quiet, peaceful, "aristocratic little city" was about to be altered," according to Dr. Tom Scott of Kennesaw State University. Some Marietta businessmen and politicians, however, saw a great opportunity to help the entire area, and the war effort. They were able to use their influence to obtain the decision to locate the new secret bomber plant just south of Marietta, on the side closest to Atlanta.

Rickenbacker was president of Eastern Airlines, and he promised that Marietta would be placed on the itinerary. He was our "Ace of Aces" in WW I, having shot down 26 German aircraft.

Mayor L. M. "Rip" Blair and the City Attorney,

James V. "Jimmie" Carmichael saw the need for the construction of this major airport. They contacted Col. Lucius Clay in Washington DC, an old influential friend, and a native Mariettan. Col. Clay, later a four-star General, head of the Berlin Airlift, and head of the Interstate Highway system, was working in Washington, DC in the Army Corps of Engineers. He was responsible for the construction of 197 new and the upgrading of 277 airports nationwide, including National Airport in Washington.

Col. Clay made sure that one of the new airports would be in his hometown, Marietta. He quietly let it be known that if Cobb would build the airport and a highway from Atlanta, he would try to make sure a major defense plant would be sent the county's way. Clay was as good as his word," wrote Joe Kirby in Marietta Daily Journal, Dec 26, 1999.

The main backers of this tremendous endeavor were: Mayor L. M. "Rip" Blair; James V."Jimmie" Carmichael, County Commissioner George McMillan, Ordinary J.J. Daniell, Congressmen Carl Vinson and Malcolm Tarver, Sen. Richard Russell, Sen. Walter F. George, Atlanta Federal Reserve Bank Chairman Frank Neely, Ivan Allen and Frank Shaw of Atlanta Chamber of Commerce, and Atlanta Mayor Buck LeCraw. Col. Lucius Clay was a crucial figure in getting Bell Aircraft in Marietta. It transformed the county. "First he got us an airport, then he got us Bell. The population nearly doubled overnight," according to Dr. Tom Scott.

Lucius Clay was recently announced to be the "Cobb County Man of the Century." He died in 1978.

Mr. Jimmie Carmichael was the key man. He grew up south of Marietta in the Oakdale community. He commuted by streetcar to Marietta High School. He was crippled as a result of being struck by a car. He graduated from Emory Law School, went into practice with Mayor "Rip" Blair, and was the Marietta City Attorney. He was also a state representative.

In the quest for the bomber plant, Carmichael made a favorable impression on Larry Bell, who made him Assistant Plant Manager when it opened. He soon became General Manager and did a superb job until it closed. He became president of Scripto Co. and ran for Governor against Herman Talmadge. He won the popular vote, but lost the county-unit vote. When Lockheed re-opened the plant, he was again General Manager. He became a

Mariettan Maj. Jack Millar   **Larry Bell**
Jimmie Carmichael
from Jack Millar

Lockheed director. After a year, he retired from the plant and was succeeded by Dan Haughton.

These community leaders united to bring Marietta a major airport and a major defense plant which built 668 "Battleships of the Air," making a major contribution toward winning the war. The plant had an annual payroll of $60 million, and changed the economic environment of Cobb County forever. Columnist Ralph McGill concluded in the "Atlanta Journal" of Feb. *25*, 1942 "that McMillan, Blair, and Carmichael had done one of the finest jobs in behalf of any county in Georgia. And in doing it, they did a job for the whole state."

After Marietta was selected for the site of the bomber plant, the next decision for the Pentagon was to decide on which corporation would build this Boeing airplane. Super-salesman Larry Bell had aggressively built the Bell Aircraft Corp. into a major company, with the P-39 Airacobra, and the P-63 Kingcobra. The Chiefs of Staff picked Bell Aircraft Corporation.

Building the Bomber Plant

No costs were spared. Robert and Company built the huge bomber plant, Air Force Plant # 6, for the Army Corps of Engineers and Bell Aircraft. The B-29 program also involved supplying crews, lengthening and strengthening runways, building

employee training facilities, etc. There were hundreds of subcontractors. This was a huge and difficult undertaking, completed in less than one year, under adverse conditions (such as hilly terrain, all kinds of weather, and wartime shortages of labor, materials, and transportation). About 700 management-level personnel were brought in from Buffalo, N.Y., home of Bell Aircraft.

Construction was started in Feb. 1942. The cost of land, buildings, and equipment was $52,300,000. There is a total land area of 2,830 acres, consisting of 150 parcels. The largest land-holders were: Cobb County, Glover Machine Works, Jordan Garner, J.J. Thomas, Charlie Thomas, T.J. Eubanks, Martin Amorous, John Fowler, Dwight Vaughn, A.S. Chapman, Frances Stevens, W.P. Kerley, Florence Douda, Harry Morrill, Jr., Amanda Chaffin, Herman Thom, Grady Kelly, Alice Mayes, George Babb, G.H. Howard, Ora Fortner, Ab Mayes, A.N. Haney, Wesley and William Childress, J.F. Hicks, George Shipp, Annie Wylie, Abbie McInnes, and many others. The Jordan Gardner home served as the Officers Club for the Dobbins ARB for many years.

The land was purchased for around $175 per acre, in contrast to $50 -75,000 for undeveloped land in that area now. There were 8,350,000 cubic yards of dirt moved, some of which was moved by horses when mud halted the bulldozers. The parking lot held 4,000 vehicles.

The B-1 Building (Assembly) has 3,194,000 square feet of floor space. The B-2 Building (Administration) has 217,500 square feet, the B3 Building (Modification) has 73,441 square feet, the B-4 Building (Preflight) has 184,000 square feet, the hangars have 197,210 square feet, the Storage Building has 99,510 square feet, the other buildings have 256,312 square feet, for a total of 4,234,000 square feet. The B-1 Building required 28,113 tons of steel, 39 miles of crane tracks, 32 miles of catwalks, 56 miles of pipe, and 31 miles of fluorescent lights. The 73 acre building had the largest number of cubic feet of any industrial building in the nation.

A railroad spur was built into it, and the streetcar from Atlanta to Marietta made a stop there. A new four-lane highway (Hwy 41), the first in Georgia, was built from Atlanta to the Canton Road underpass, and the Access Highway, now South Cobb Drive, was built.

The airport was newly finished when the first B-29 rolled off the assembly line on November 4, 1943.

The Influx of Workers

In 1940, the Cobb County population was 38,000, most living on modest farms with about 78% sharecroppers. Marietta was the county seat, a charming and proud small town with a population around 7,000 people. Nearby were the small towns of Kennesaw, Smyrna, Austell, Acworth, Powder Springs, and Mableton. (At the present time, Cobb County's population is over 530,000, and it has become part of Metropolitan-Atlanta.)

At that time, Marietta's economy depended primarily on agriculture, especially cotton. It also had the Glover Machine Works (which made small locomotives), the McNeel Marble works (which made tombstones), the Brumby Chair Factory, Holeproof Hosiery, and several small textile factories. A common wage was under ten cents per hour. Glover Machine Works paid thirty five cents, and Bell Aircraft paid eighty nine cents per hour. These survived the Great Depression, and softened the hardship on the area. It was a religious and cultural town with a great deal of civic pride. There are still many antebellum homes. There were some tourists.

Marietta's population tripled, and Cobb county doubled during the war. When the bomber plant was ready to begin production, Marietta's unemployment problem became a labor shortage, especially for the B-29 assembly line. Skilled workers were few. Most able-bodied men were in uniform. To fill these jobs, workers came from all over Georgia, especially from the mountains. The "red blood' from the mountains and the "blue blood" from old Marietta mixed nicely with the "Yankee Blood" from Buffalo, N.Y.

Employees were recruited from all over the southeast. Eighty two percent came from Georgia and 11% came from surrounding states. Many commuted daily from Alabama, Tennessee, and north Georgia. Workers were attracted by the high wages of 89 cents per hour, and there was a lot of overtime. Many workers also came for patriotic reasons, mainly to do their bit toward winning the war.

Marietta accepted the tremendous influx of strangers and Yankees very well. Basically these outsiders were integrated smoothly into the churches, businesses, and communities. There was a surplus of humor regarding Yankees, Damn Yankees, Southern accents, and customs. Some Southern employees were not always pleased

about having to take orders from these Yankees. My mother-in-law was described as being a very nice lady, even though she was a Yankee from New Jersey. Most of the 700 management people who were moved from Buffalo elected to remain in the Marietta area after the war.

Marietta Place

Housing the war-workers who came to Marietta was one of the biggest problems facing the town. Tiny Marietta was swamped with new-comers. The city had recently completed the construction of the Clay Homes for low income families. Now the government built the huge Marietta Place apartments. One thousand units of temporary construction were built on the site of the county fairgrounds and the city prison. These Marietta Place "temporary" apartments were finally torn down in 1960. The county built a stairway up from Marietta Place, across the highway, approaching the front gate of the plant. These steps are now preserved as the Bell Bomber Park. They permitted the war-workers to walk to work, conserving rationed gasoline, saving parking, and possibly avoiding the need to purchase a car. Detroit had stopped making new cars - busy building Jeeps, trucks, tanks, landing craft, and airplanes to fight the war.

Other housing projects consisted of the 500 apartment units of Pine Forest, Boston Homes, and duplexes in the Hedges-Gramling-West Dixie areas. All these units were still not adequate, and many "Old Mariettans" graciously opened their homes or made small apartments out of their garages or spare space. For example, a chicken house was converted into two apartments. Most Mariettans endured inconvenience and upheaval for patriotic and other reasons.

The federal government and the City built a small shopping and entertainment center called "Larry Bell Center" near Marietta Place at Clay Street and Fairground Street. This consisted of a gymnasium, a swimming pool, bowling alleys, a theater, Dunaway's Drug Store, a grocery store, small shops, and Dr. Bruce Burleigh's office. A fire destroyed the gym and bowling alleys about 1960. This property is now the site of the Romeo Hudgins and Jenny Anderson complex, and an aquatic center.

Bill Kinney in the Oct. 2, 1999, <u>Marietta Daily Journal</u> writes:

"The Federal Government in 1942 built a 1,000

Larry Bell Park recreation center
Robert Crowe collection

Bell Bomber Park dedication 1996
Mayor Ansley Meaders in white

Bell Bomber Park Stairs
From Marietta Place to Bell Bomber Plant

apartment 'temporary housing' project on Fairground Street across from the present site of the Marietta Daily Journal to house workers at the Bell Bomber Plant, now Lockheed-Martin. Workers could walk up the high set of steps in the current Bell Bomber Park and report to work assembling B-29 Superfortresses whose bombs would turn Japanese cities into ash heaps.

The area had its Bell theater, gym, a swimming pool, a bowling alley, a drug store, restaurant, and other businesses. Marietta Place residents were closely knit.

When Japan surrendered in August 1945, many thought Marietta Place would be razed immediately. Not so. The Feds gave the City of Marietta the apartments and they continued for a number of years afterward.

Each year, Marietta Place residents and those who lived in other single family residences hold a reunion at the American Legion Post on nearby Gresham Street. Richard Culver who yearly shepherds the group reminds the party starts today at 10:00."

In addition to the housing for this influx of workers, churches and schools were built or expanded. Other services were stretched to meet the needs of the people. There were two newspapers: The Marietta Daily Journal, and the Cobb County Times (a weekly). The old privately-owned Marietta Hospital in downtown Marietta was enlarged to 50 beds. The few doctors served Marietta well during some very trying times when facilities and staff were sorely taxed.

The USO (United Service Organization) was very active every week-end in the effort to welcome "GIs" and make them feel at home. Dances were held on the second floor of the Marietta City Hall, south of the old Post Office, under the leadership of Regina Rambo Benson. Several good marriages resulted. Many other groups exhibited Southern Hospitality.

Training the new employees was a major job. Many came from out of the hills and rural areas with no industrial skills, and sometimes illiteracy was a problem. New-hires were trained in Buffalo, Atlanta, Fulton County, vocational schools in Cobb County, and at the plant. In Marietta the former CCC (Civilian Conservation Corps) and NYA (National Youth Administration) facilities were used as training schools. The NYA building on Roswell Rd. was used for training, and was called

Rickenbacker School, thanks to a $35,000 gift from "Captain Eddie."

Mrs. Dempsey (Corene) Kirk, from Marietta, states that she started employment at Bell with an hourly wage of $0.75 and worked up to $1.25, with time and a half for overtime, and double time for weekends. She frequently worked 83 hours per week, on the day shift, as an electrical assembler. She and her sister were able to buy a large farm during that period. She states morale was high and the employees "felt like family."

Bell employee, Ruth Scarr Inglis, 1944

My wife, Ruth Scarr Inglis, was a Bell (not Bell South) summer employee between semesters in college in 1944, as a clerical worker, for $0.85 per hour. Her father, Francis J. Scarr, Sr., was an engineer at Bell for about two years, having come from Curtiss-Wright in St. Louis, and Grumman Aircraft on Long Island. Her family found private housing on three occasions, one of which was sharing a home with a generous Mariettan.

Many businesses were started and thrived during this era. The entire area grew rapidly. Federal money aided all the local governments through "impact grants" and funds to help with development of schools, housing, streets, and infrastructure. A Hill-Burton Act Hospital was started, but then stood incomplete for four years until 1950. The new 100-bed hospital was dedicated in July 1950 and was named "Kennestone Hospital." The name Kennestone was derived from the fact that Kennesaw Mountain and Stone Mountain were both visible from the hilltop overlooking the entire area on Church Street north of town.

Marietta never did return to its pre-war status, but continued to boom from the momentum gained from Bell Aircraft. The small town became a cosmopolitan city, and continues to thrive today, in part due to the impact and impetus of B-29s like *Sweet Eloise.* The era of the B-29 had begun, and continues to this day, a fact not appreciated by many Mariettans. Marietta can be proud of its major contribution toward the final Victory on August 15, 1945.

Walter (Beau) Clark, a Lockheed retiree, appreciates the B-29 for four reasons: (1) they probably saved his life when he was in the 3rd Marine Division on Guam scheduled for the invasion of Japan on November 1, and would probably have been killed or wounded in his assignment. The B-29 obviated this invasion; (2) He has had excellent employment at Lockheed in the old B-29 plant; (3) The B-29 gave impetus to making Marietta a fine place in which to live and raise a family; and (4) The B-29 contributed greatly to the defeat of the Axis, thereby giving us this great country in which we live, and making us a superpower. The B-29 is part of Tom Brokaw's Greatest Generation.

Statistical Summary Sheet of the Bell Aircraft Plant

The peak week at Bell Aircraft was the week ending February 18, 1945. There were 28,280 people employed: 17,867 males, 10,413 females. There were 2,264 "colored."

There were 546 new employees of whom 330 were trainees in that week There were 446 separations, of whom 14 joined the armed forces, either voluntarily or by draft. There were 4,733 vulnerable to the draft. There were 9.5% absentees, which was approximately 3,000 persons.

The average hourly base pay was $0.89, earned was $1.01, for 40 hours. The weekly base pay was $55.00, earned was $83. The average salaried weekly pay was $74, earned was $98.

Bell Aircraft Production Lines
Bill Kinney Collection

STATISTICAL SUMMARY

WEEK ENDING __2-18-45__

GEORGIA DIVISION OF
BELL AIRCRAFT CORP.

PRODUCTION

UNIT	THIS WEEK SCHED	COMP	TO DATE SCHED	COMP	OVER	UNDER	LAST SHIP IN FIXTURE OR DEPT.
DEPARTMENT 38							
Rear Beam	14	13	418	436	18		441
Front Beam	14	16	406	423	17		427
Trailing Edge	14	14	404	419	15		426
Beam to Beam	15	13	378	389	11		407
DEPARTMENT 42							
Wing Installations	15	15	364	362		2	374
DEPARTMENT 52							
Front Fuselage—Sec. 41	16	14	397	381		16	396
Center Fuselage—Sec. 42	13	12	390	392	2		404
Wing Gap—Sec. 43	12	14	394	401	7		417
Rear Fuse. Fwd.—Sec. 44A	13	11	383	371		12	391
Rear Fuse. Aft.—Sec. 44B	10	11	400	407	7		425
Tail Cone—Sec. 45	14	13	392	396	4		406
DEPARTMENT 86							
Leading Edge	14	28	425	405	--	20	453
Tail Cone—Sec. 45	TRANSFERRED TO DEPT. 52						
DEPARTMENT 53							
Nose Section—41	15	12	364	356		8	380
Center Fuselage—42	15	14	364	363		1	379
Wing Gap—43	14	14	365	374	9		398
Fwd. Sec. Rear Fuse.—44A							359
Aft Sec. Rear Fuse.—44B	15	10	363	362		1	369
Tail Cone—45	15	13	371	375	4		395

*Joined.

BANK—READY FOR DEPARTMENT 63

SHIP SECTIONS	11	41	42	43	EMPENNAGE
QUANTITY		3	--	14	15	7

DEPARTMENT 63

ASSEMBLY LINES B-1 BUILDING (1) SOUTH/NORTH STA. NO.	FINAL MOD.	PRE-FLIGHT LINE	BELL FLIGHT LINE	B-8 INSP.	ARMY FL'GHT	ACCEPTED BY AAFRR	DELIVERED ATL
	661						
	712 711	662 641					
1	707 710	663 645	624 635	627	609	609
2	706 705	644 633	621	619	619
3	704 —	661 643 634	628	623	623
4	702 703	649 632	630	618	618
5	— 701	650 639	622	620	614
6	700 699	671 651 642	626	614	630
7	698 697	672 652 640	617	630
8	696 —	673 653 641		
9	691 695	674 654 631		
10	694 693	676 655 643		
11	690 692	677 656 625		
12	689 687	681 657 637		333
13	688 —	678 658 638		
14	685 686 680 659 629		
15	684 683 660 647		13

(A) AIRPLANE STILL ASSIGNED TO PREVIOUS ACTIVITY.
REWORKS

(1) CUMULATIVE PRODUCTION NUMBERS CONVERT TO AAF AIRPLANE NUMBERS AT STATION 1, DEPT. 63.

PRODUCTION FLOW

PRODUCTION—COMPLETION STATUS

DESCRIPTION	THIS WEEK SCHEDUL'D	PRODUCED	TO DATE SCHEDUL'D	PRODUCED	OVER	UNDER
Overrun Ships	xxx	xxx	5	5	xxx	xxx
Bell Ships DEL. TO ATC	12	6	276	268		8
Equivalent Ships—Bell	11	13.1	415	399.7		15.3
Complete Spar Chords	22	56	914	834		70
Equivalent Spar Chords	22	29.9	914	924	10	
Spares—Thousands of $	NOT AVAILABLE					

EQUIVALENT AIRPLANES AND AVERAGE HOURS
FOR WEEK ENDING 2-18-45

	EQUIVALENT SHIPS	ESTIMATED HOURS	ACTUAL HOURS
This Week—Bell	13.1	70,000	48,680
To Date—Bell	399.7	80,000	72,776

MANPOWER

LABOR GROUPS	MALE	FEMALE	TOTAL ACTUAL	PER CENT	SCHEDULED	COLORED
Direct Employees	8725	5504	14229	50.1	16460	984
Indirect Employees — Plant	8468	4368	12836	45.	xxx	1273
Indirect Employees — Outside	39	16	55	—	xxx	xxx
Indirect Employees — Trainees	635	525	1160	4.	xxx	7
Total Indirect Employees	9142	4909	14051	50.	13440	1280
Total Employees	17867	10413	28280	100	29900	2264
Per Cent of Total Employees	63.	37.	100.	xx	xxx	8.
Contractors Employees in Plant on Bell Work	xxx	xxx	3	xx	xxx	xxx
GRAND TOTAL	xxx	xxx	28283	xx	xxx	xxx

AVERAGE WAGE RATES
FOR WEEK ENDING 2-11-45

		Base		Earned	
Hourly		Base	$.89	Earned	$1.01
Weekly		Base (40 Hours)	$55.	Earned	$53.
Straight Time Weekly		Base (40 Hours)	$74.	Earned	$93.

EMPLOYMENT

	ACTUAL	SCHEDULED
Hired This Week—Direct	100	(B) 328
Indirect (330 Trainees)	446	#
Total	546	328
Separations This Week—Armed Forces	14	xxx
Other	432	xxx
Total	446	xxx
Labor Turnover—This Week	1.65%	xxx
Employees Vulnerable To Draft — Ages 18-25	80	xxx
Employees Vulnerable To Draft — Ages 26-29	1061	xxx
Employees Vulnerable To Draft — Ages 30-37	5733	xxx
Absentees—Daily Average of All Employees	9.5%	xxx

PLANT STATISTICS

PROJECT STARTED MARCH 30, 1942 — AIRPORT ACQUIRED MAY 9, 1944
First Steel Erected B-1 Building August 31, 1942.
Cost of Land, Buildings and Equipment $52,300,000

BUILDINGS
B-1 Main Assembly Bldg., First Floor	
Floor and Mezzanines	2,488,151 sq. ft.
B-1 Basement	706,700 sq. ft.
B-2 Administration	217,488 sq. ft.
B-3 Final Modification	73,441 sq. ft.
B-4 Pre-Flight Hangar	194,680 sq. ft.
Airport—Hangars	197,210 sq. ft.
Storage and Warehouses	99,510 sq. ft.
All Other Buildings	257,312 sq. ft.
Total Floor Area All Bldgs.	4,234,492 sq. ft.

LAND & MISCELLANEOUS
Industrial Area	429 acres
Airport Area	1220 acres
Total Area	2830 acres
B-1 Main Assy. Bldg.	2000 ft. X 1024 ft.
B-1 Main Assy. Bldg.	28,113 tons steel
Crane Tracks	39 miles
Catwalks	32 miles
Water and Steam Piping	56 miles
Parking Space	4000 autos
Earth Moved (Original)	8,350,000 cu. yds.

Air Conditioning 7,200,000 cu. ft. per minute.
Landing Runways, three each 6,000 ft. long X 150 ft. wide.

(B) INCLUDING PREVIOUS DELINQUENCIES.
INDIRECT HIRES FOR REPLACEMENT ONLY.

Bell Statistical Summary, peak week

Bell Flight Line 1945
Photo: Courtesy Smyrna Historical and Genealogical Society and Museum

There were thirteen B-29s delivered and accepted that peak week, with 48,680 man-hours expended per ship. Production efficiency was averaging 72,776 man-hours per aircraft for the first 399 B-29s produced, and some employees were able to work a normal 40 hour week when production got rolling. Bell produced four B-29s in 1943, 201 in 1944, and 463 in 1945. During 1945 they averaged about 60 per month during peak production, while Wichita averaged about 100 ships, Renton about 140, and Omaha about 55. Renton broke all records with 160 during the month of July. Frequently B-29s were flown from Wichita to Marietta to exchange know-how. The workers were represented by the United Auto Workers Union.

There were 19 production stations on the B-29 assembly line, consisting of Rear Beam, Front Beam, Trailing Edge, Beam to Beam, Wing Installation, Front Fuselage, Center Fuselage, Wing Gap, Rear Fuselage Forward, Rear Fuselage Aft, Tail Cone, Leading Edge, Tail Cone, Nose Section, Center Fuselage, Wing Gap, Rear Fuselage Forward, Rear Fuselage Aft, Tail Cone.

For a good perspective of the Marietta war time feeling, the 20-minute documentary movie by the War Department and Bell Aircraft is good viewing,

and is available at the Smithsonian National Air and Space Musuem, entitled "B-29s Over Dixie."

Production

The Marietta bomber plant produced 668 B-29s, of which 357 were B-29A models, and 311 were B-29B models. The Bs were the stripped down versions, and were considerably faster, able to fly higher, carry a heavier load, and with a longer range than the A models. Gen. LeMay requested a lighter and faster version of the B-29 for incendiary and night-time bombing, since the Japanese fighter opposition was far less than expected, and to minimize the threat from anti-aircraft fire. Marietta was the main producer of the B Model. The stripped-down versions were also used as weather and reconnaissance planes. They did not have turrets, machine guns, or armor plate, and had a smaller crew. They carried only two machine guns, located in the tail. Some B models weighed 7,000 pounds less than an A model, increasing the speed 10 mph, with a maximum speed of over 365 mph.

With so many aircraft being produced and tested, Marietta became a very noisy place at times. A lady in Smyrna complained that when the props of some

THE WEATHER

Yesterday: High, 45. Low, 25.

Today: Partly cloudy and warmer.

THE ATLA

For 78 Years an Inde

VOL. LXXVIII., No. 223.

ATL

B-29 TAKES ROAD TRIP—A gift of the Bell Aircraft Corporation to the city of Marietta, this B-29 Superfortress must be moved from the bomber plant to the Larry Bell recreation park, a half-mile away. A portion of the fence is being removed here as the bomber starts on its last journey to become a monument to war work.

CAREFUL, NOW, THAT IT DOESN'T TAKE OFF—It's quite a transportation feat moving the huge ship, for its made for flying. A tractor eases the craft down the incline toward the highway. The bomber is a gift from Larry Bell to the city of Marietta.

ON THE HIGHWAY AND DOWN THE MIDDLE—Portions of the wingspread had to be cut off so that the ship could make its highway journey, and even in it's abbreviated state trees had to be felled along the route to allow passage. The bomber will memorialize the city of Marietta's superb job of producing the big planes for war on Japan.

A ROAD HOG IF THERE EVER WAS ONE—The ship is towed down the highway and a caravan of trucks and automobiles follow. But there's room for only one-way travel. The Larry Bell Park is a $200,000 recreation center for Marietta and Cobb county.

Surplus B-29 being moved to Fairground Street January 26, 1946 Atlanta Jounal & Constitution from Bo Glover

**Surplus B-29 on Fairground St. Site of Perry Parham Field
from Mrs. Robert (Ellen) Blackwell**

planes were "out of sync," the dishes rattled so much they occasionally fell off the shelves and broke. There was a nearly constant roar from engines being tested on the test stands, day and night.

Dan Cox, curator of the Marietta Museum of History, claims that as a boy, he and his buddy, Don Donley, would shoot bows and arrows at the awesome low-flying giants. He claims they hit one. They were terrified that it was going to crash. They ran into the house in case it did crash and they would be blamed. Dan states that they sometimes flew so low that the branches of trees would wave. They flew much lower, slower, and were louder than today's jets. Dan tells of huge flocks of starlings nesting in the trees on the Square. He said low-flying B-29s were sent over the square to disturb the birds, and black clouds of noisy birds would take off. The starlings have since disappeared. In the hot weather, with no air conditioners and wide-open windows, the continuous noise from the '29s was aggravating.

Jimmie Carmichael and Larry Bell introduced many progressive programs, such as child care, car pooling, blood drives, war bond drives (enough to purchase a new B-29), and income tax deductions. There were 1,757 handicapped employees, including the blind and the deaf, who made excellent workers and they caused few problems. Minorities and women were given increasingly responsible and difficult jobs.

After a slow start on the production line, the production efficiency progressed to around the rate of 13 ships per week on the twin assembly lines which was considered an outstanding rate.

Fairground Street B-29

At the end of the war there were 32 unfinished ships at various stages of completion. One B-29, without engines, was donated by Bell Aircraft to the City of Marietta as a memorial. With considerable difficulty it was moved down the steep hill to Fairground Street, after trimming the

wing tips, and removing some trees and poles. It was parked across from Marietta Place where Perry Parham Field is now located. All the kids from the area played on it and in it, enjoying the cockpit and tunnel the most. It was badly beaten up and it suffered some storm damage. "Wild Man' Dent Myers states he assisted in scavenging parts to be used in Lockheed's B-29 renovation program. To most Mariettans, it gradually became a sad relic of the war, and most everybody was glad when it disappeared, as a result of being scrapped, in 1952.

Victory

With the arrival of VJ Day (Victory Over Japan) there were mixed emotions in Marietta. People hated to lose an excellent job, and yet they were jubilant that this country was the victor in a terrible war. They knew that their jobs were winding down, and many had already left their wartime jobs. Most had no trouble finding work, but they were concerned about competition from other wartime workers and by millions of returning veterans. These returning vets proudly wore their "Ruptured Duck" lapel pins. Most of the veterans were given preference in employment over non-veterans.

The people of Marietta never had it so good. "How you gonna keep 'em down on the farm, after they've seen Paree?" was a popular song after WW I, but it was applicable to this war, also. People had money in their pockets, but due to scarcities there was not a lot to spend it on. Rationing was now a thing of the past. Price controls had done a good job during the war. The Depression was forgotten. There was prosperity. Women and minorities had been empowered. Victory celebrations were subdued, and there was a genuine sense of relief and pride However, there was concern about impending inflation, about the future, about Communism, about the devastation in Europe and Japan, and about the horrors of the atom bomb and the atomic age.

When the war ended about 12,000 workers were rapidly terminated. The closure of the "bummer plant" was well accepted by the employees. There was speculation about the future of the vacant plant. Marietta did not become a ghost town. Marietta Place stayed full and employment was high because of the prosperity, the growth of Atlanta, and the pent-up demand for everything.

Marietta Place rapidly filled up with students who were taking advantage of the "G.I. Bill."

Lockheed

When the unfinished B-29s were moved out, the Air Force Plant #6 soon almost filled up with valuable and expensive machine tools from the war production around the south-east. This was handled by the Tumpane Company. In 1951 the plant was re-opened by Lockheed for the Korean War to rehab 175 B-29s, and to start production on the Boeing six engine turbo-jet bomber, the B-47, to succeed the B-29. The B-47 was also produced in Wichita. Lockheed produced 394 of the B-47s, followed by the C-130, the C-141, the C-5A and the C-5B, as well as others. During the war Marietta was known as the "Bomber City," and then became known as the "Air-lift Capital of the World." Lockheed is still building C-130s after more than 47 years, which is the longest production run in history.

The B-29 program had been a tremendous gamble, both militarily and financially. It was a three billion-dollar gamble, and the atom bomb was a two and a half billion dollar gamble. They were both risky, but they paid off. The man-hours, the money, and the sacrifice, were justified by their success. The New York Times, December 11, 1944 states that the B-29 project was the greatest single undertaking in industrial history.

According to Jacob Meulen in Building the B-29, the entire B-29 project cost $30 billion, in 1990 dollars, when one calculates the plants, the aircraft, the parts, air bases, training, personnel, fuel, bombs, etc, for about 16 years. The original cost was around $650,000 per unit. This money was crucial in waging two wars plus the Cold War. It corrected economic problems and prompted social change all the while providing national security.

The war caused some major socio-economic changes in the status of women and minorities as the walls of discrimination were beginning to crumble. Most of the women were not "Rosie The Riveter," but many were. They performed many of the "masculine" jobs with excellent results. Many elected to stay in the work force. Most were clerical workers, including Mrs. Eloise Short Strom, who later lent her name to our B-29 *Sweet Eloise*. She, along with millions of other women,

had a war time job, and elected to remain employed of her own volition after the war. She is still employed at Rich's Department Store.

During the war, there were three B-29s that had to make emergency landings in Russia at various times. Russia refused to return these aircraft, despite the fact that we were allies. They proceeded to make precise copies of our B-29 and put them in production. They eventually built over 1,200 of these replicas. We even supplied them with brakes and tires. These clones were then upgraded, as we did with the B-50. They were known as the Tu-4, the Tu-80, and the Bull. They comprised the atom-bomb carriers for Russia for several years.

When production of B-29s ended, Wichita had built 1,644, Renton had built 1,119, Marietta had built 668, Omaha had built 536, and Seattle had built three (which may have been prototypes), a total of 3,976 B-29s. For comparison, Consolidated built 19,203 B-24s; North American built 15,586 P-51 Mustangs; Boeing built 12,731 B-17s; Lockheed built 9,942 P-38s; Messerschmidt built 35,500 ME 109s for Germany; Russia built 35,143 Ilyushin IIs; England built 20,351 Spitfires; and Mitsubishi built 10,499 Zeros for Japan.

The U.S. had built 299,000 warplanes during the war. After the war, the tremendous surplus of war material created problems, solved by being sold or scrapped or stored. A war-weary B-29 which had cost $665,000 could be bought for $350 for museum purposes, and somewhat more for scrapping and other purposes. When the war ended, a total of 5,092 orders for the amazing B-29s had to be cancelled nationwide.

Chapter 13 ————
Marietta s B-29 Heroes ————

- Jack Millar
- Ed Wigley
- Gilbert Johnson
- Joel Rutledge

- John (Bo) Kincaid
- Dempsey Kirk
- John Frey
- Bill Price

Marietta had seven known "native sons" who flew B-29s during the war, all of whom survived. They all made a contribution.

Jack Millar, Col. was the first. He graduated from Marietta High School in 1936, where he was an athlete and a scholar. He planned on being a reporter for the <u>Marietta Daily Journal</u>. He graduated from the Citadel in 1939.

Jack joined the Army Air Corps and became a B-17 pilot. He participated in key moments of the war. He was a B-17 pilot in the first bombing of Germany in 1942. He delivered top-secret battle plans to US military leaders from Gen. Eisenhower. He then became a B-29 pilot and took delivery of the 10th B-29 off the Bell assembly line, which he christened the *Georgia Peach.*

In 1944 Jack was the pilot of *Georgia Peach* on the first bombing raid of Japan since the 1942 B-25 Jimmie Doolittle raid, in which they hit the Yawata Steel Works. Thus Jack has the distinction of being the only pilot to fly a B-17 on the first raid on Germany, plus the first raid on Japan in his B-29. This was also the longest bombing raid in history at that time. Jack flew two other bombing missions against the Japanese Empire. He was stationed in the CBI (China-Burma-India) theater with the 58th Wing and flew 26 "Hump" missions over the Himalaya Mountains. He was proud of his Marietta *Georgia Peach.*

Jack returned to the States, and continued to contribute to the Army Air Corps in many ways, including instruction of Paul Tibbets, who later made his historical flight over Hiroshima. He wrote the Air Force Instrument Manual, and he became Chief of all B-29 training. He assisted in writng the plans for the invasion of Japan. He

Georgia Peach **and Maj. Jack Millar   from Jack Millar**

retired from the Air Force as a Lt. Col. and is partially disabled as a result of a B-17 crash which was caused by lightening resulting from the Alamogorda atomic bomb testing. Jack now resides in Winston-Salem, N.C. He is a candidate for the Georgia Aviation Hall of Fame which is located at Warner Robins AFB. He has a recommendation from Captain Eddie Rickenbackeer. Jack and his *Georgia Peach* are featured in the January, 2001 issue of World War II magazine entitled "First Superfortress Attack on Japan."

Ed Wigley, T/Sgt, graduated from Acworth High School in 1940 and went to work for Western Auto located on the Marietta Square and again after the war. He then went to work for Lockheed from which he retired. Ed presently resides in Marietta.

Ed was a radio-radar operator in the 58th Bomb Wing and the 468th Group. The radar men were also trained as navigators, bombardiers, and First-Aid men for the ship. He was in the CBI theater. He operated from Kharagapour near Calcutta and he flew out of Chengtu, China, on 25 bombing missions. These included two mining missions of Shanghai and one of Singapore in which they succeeded in sinking a massive dry-dock. He made 11 logistic flights over the "Hump." Mining (dropping one-ton aerial mines from a low altitude) was extremely effective. All told, B-29s dropped 12,000 of them which almost completely blockaded the Japanese islands, and sank 1,250,000 tons of shipping, as much as the Navy and submarines did in the final five months of the war.

Ed's Wing was then moved to Tinian Island where he flew another 10 missions over Japan. Four of these missions were on board *Jolly Roger*, surviving severe flac damage, but Ed never needed his First Aid skills.

Gilbert Johnson, S/Sgt. graduated from Robert L. Osborne High School in 1941. He was in the service from 1943 to 1946, where he served as a B-29 radio-radar operator as well as First Aid / navigator/ bombardier. He was in the 315th Bomb Wing and 331st Group on Guam. He flew in a Marietta-built ship, and states that all the B-29s in his 315th Wing were built in Marietta.

Gilbert volunteered late in the war, and he received credit for two missions which consisted of two Mercy drops on American POW camps after the surrender. These consisted of 12,000 pounds of food and supplies. This was in 300 lb. cartons strapped on pallets, each with a parachute. Sometimes 55 gallon drums were used. There was considerable risk of killing several POWs with these drops. His Group lost two B-29s from crashes during the POW Mercy drops. One of Gilbert's drops was on POW Camp Omori located near Tokyo, which is the camp in which Bill Price from Marietta was a prisoner.

After the war, Gilbert returned to work at the Anderson Chevrolet Co. where he retired as Sales Manager after 46 1/2 years. He presently resides in Powder Springs.

Joel Rutledge, S/Sgt, graduated from Kennesaw High School in 1935, and presently resides in Marietta. He served in the Army Air Corps from 1943 to 1945 as a tail gunner. He was in the 73rd Bomb Wing and the 498th Group on Saipan, almost adjacent to *Sweet Eloise*. They carried a big "T" on the rudder signifying the 498th Group instead of the big "Z" of the 500th Group. Most of his missions were in *Little Joe*.

They received flac or bullet damage on most of their missions, but no injuries. Two pieces of shrapnel perforated his chest parachute, but didn't scratch him. In another miraculous incident, while taking off in his heavily loaded aircraft the propellers touched the surface of the water. They managed to return to the airbase, but two engines and propellers had to be replaced.

John (Bo) Kincaid, T/Sgt, DFC graduated from Marietta High School and attended Young Harris College. His was a prominent pioneer family which donated the land for Kincaid Elementary School in East Cobb County. After the war he was employed by his family's seed and feed business on Powder Springs Street. He was killed in a car accident in 1948.

He witnessed the atom bomb on Hiroshima, as one of two escort and observer ships. He was an eyewitness to "Doomsday," and saw the mushroom cloud. Undoubtedly, some of the photographs of the atomic bomb that we see were taken from "Bo's" ship. His aircraft was built in Marietta, which he described as a "magnificent ship." He flew 23 missions, including one mining mission, more than any other Mariettan in a Marietta-built ship. He was a gunner. He flew with the 73rd Bomb Wing on Saipan and the 509th Composite Group out of Tinian Island.

Dempsey Kirk, T/Sgt, graduated from Acworth

High School in 1937. Dempsey has always lived at the foot of Kennesaw Mountain in Marietta. After the war he was a ranger at Kennesaw Mountain National Battlefield Park, then he retired as an electrical technician at Lockheed.

He was in the Army Air Corps from 1942 - 45. He was stationed on Saipan with the 73rd Wing and 497th Group as a gunner. He flew 12 missions. His aircraft touched the water one time on takeoff, with no damage. He had two blowouts on takeoff. He had two emergency landings on Iwo Jima, one for fuel and one for a runaway prop.

John Frey, Jr., 1st Lt. graduated from Marietta High in 1941, then attended Presbyterian College. He volunteered in 1944 for the Army Air Corps and became a B-17 pilot. As victory in Europe came closer, B-l7 pilots and crews were assigned to B-29s. John completed his B-29 pilot training in Albuquerque, New Mexico and was ready to go into combat when the war ended.

After the war, he attended University of Georgia. John then operated the Frey Cotton Gin in Kennesaw, Georgia which later became the structure of the Civil War "General" train museum. This will soon be upgraded with a $5,000,000 expansion. He became part owner of the "Big Chicken." He became a contractor. John served on the Board of Education. He continued to fly, owning his own airplane. Jack died of a heart attack in 1990.

Bill Price, S/Sgt., Purple Heart, POW, is a Mariettan by choice since 1986. He retired from Ford Motor Co. He is a charter member of our B-29 Superfortress Association and is the treasurer. He is a director of a Georgia Prisoner-of-War organization.

Bill flew seven missions as a gunner with the 314th Wing, 29th Group out of Guam. On April 7th 1945 over Nagoya, after dropping their bombs, a Kamakazi hit his B-29. With the aid of centrifugal force he was able to break the side blister, and fell out. He has scars to show for it. He "came to" with his parachute open. Of the eleven-man crew, two others survived, and the three men were almost killed by angry civilians. Bill had multiple fractures and contracted several diseases. In the Omori POW Camp in Tokyo Bay, Bill's weight dropped from 150 lb to 103 lb. in five months, with no medical care. His experiences are written up in Robert Martindale's excellent book, 13th Mission. This is the same camp in which Mr. Hap Halloran, a well-known POW, was a prisoner.

During the war, the POWs were concerned about getting killed by the American bombs. After the war they were concerned about getting killed by falling American 55 gallon drums in Mercy drops. After the surrender, Bill's POW camp near Tokyo received two POW Mercy drops of food and supplies, one of which was definitely performed by *Sweet Eloise,* confirmed Lt. Wanless Goodson as co-pilot. The second drop was performed by Mariettan Gilbert Johnson in his Marietta-built B-29. Bill's life has been definitely affected by B-29s.

BIBLIOGRAPHY

B-29 –44-70113 by Alton Evans   Southern University Press 1999

B-29 Photo Combat Diary by Chester Marshall   Specialty Press 1996

B-29 Superfortress by Chester Marshall   Motorbooks International 1994

B-29 Superfortress in Action by Steve Birdsall   Squadron/Signal Publications 1980

Building The B-29 by Jacob Vander Muelen   Smithsonian Institution 1995

Bombers of World War II by David Donald   Metro Books 1998

Bombers Over Japan   by Keith Wheeler   Time/Life 1982

Combat Aircraft of World War II by Bill Gunston   Bookthrift Publications

Final Assault On The Rising Sun by Chester Marshall   Specialty Press 1995

Flight Of The Enola Gay by Paul Tibbets   Buckeye Aviation Book Co. 1989

Greatest Generation by Tom Brokaw   Random House 1998

Greatest Generation Speaks by Tom Brokaw   Random House 1999

History of WW II   by Col. Eddy Bauer   Gallahad Books 1979

Marietta Daily Journal,   Marietta, GA.

Sky Giants Over Japan by Chester Marshall   Global Press 1994

Smithsonain Associates   September 1995

Strategic Air War Against Japan by Maj.Gen. Haywood S. Hansell Jr.
                              Airpower Research Institute 1980

Stars and Stripes   Paris Edition Vol.1 No. 318

The Bombers by Alan Cross

The Story of Offensive Strategy and Tactics in the Twentieth Century by Robin Cross
                                          MacMillan   1987

The Fall Of Japan by Keith Wheeler   Time-Life Books 1983

Trial By Fire by Christopher Lew   World War II September 1995

Winning WW II in an Atlanta Suburb Essay by Dr. Tom Scott, Kennesaw State U. 1998

ABOUT THE AUTHOR

Dr. Pete Inglis, born and raised in Nebraska and Colorado, has always been interested in aviation and is a frustrated pilot. While attending Washington University School of Medicine in St. Louis, he was drafted into the ASTP (Army Specialized Training Program) in 1943. After graduation and an internship, he was called to active duty in 1947. He volunteered for Flight Surgeon school at Randolph Field in Texas where he also received some pilot training. He was assigned to Yokota AFB, Japan, where he was on flying status. He was required to fly as an observer for eight hours per month, sometimes in the huge B-29s. His father-in-law had been an engineer at The B-29 Bell Bomber Plant and his wife a brief Bell employee.

After his discharge in 1949 and more medical training, Dr. Pete engaged in general practice in Marietta, for 38 years. He worked briefly at Lockheed. In the late 1980s he generated local interest in obtaining a B-29 for Marietta for historical purposes. To spearhead this interest, the B-29 Superfortress Asssociation, Inc. was formed in 1991 with Dr. Pete acting as Historian for the group. The efforts of the Association to bring *Sweet Eloise* (*44-70133* or *Z-58* or *Marilyn Gay*) to Marietta was a story "waiting to be told." Dr. Inglis, now in retirement from medicine, took it upon himself to write this unique and unusual story, <u>Restored to Honor: Georgia's B-29 *Sweet Eloise*</u>.

DEDICATION

This book is dedicated
to the aircraft, B-29 *44-70113, Z-58.*
She served with honor during World War II and the Cold War,
collapsed, was rescued, rebuilt, restored, renamed
Sweet Eloise
dedicated May 6, 1997 Dobbins ARB
as a symbol of this community's war effort in the production of B-29s
at the Bell Aircraft Plant,
Marietta, Georgia
1942-1945

MW00713353

STATE OF GEORGIA
OFFICE OF THE GOVERNOR
ATLANTA 30334-0900

Roy E. Barnes
GOVERNOR

Dr. Pete Inglis
80 Lindley Ave.
Marietta, Georgia 30064

Dear Pete:

Congratulations on the publication of your book, *Restored to Honor, Georgia's B-29 "Sweet Eloise."* It is a story I can recommend to all Georgians, especially to those interested in aviation and wartime Georgia.

I would like to take this opportunity once again to congratulate the B-29 Superfortress Association, Inc. for its notable accomplishment in restoring the wreckage of *Sweet Eloise* to its present impressive condition. As you know, I was involved in this project at its inception and, as a member of the 1996 General Assembly, spearheaded the effort together with Governor Zell Miller and State Senator Chuck Clay to assist the Association in completing the restoration. It is an important piece of Georgia's history and one in which we can all take great pride.

I look forward to reading *Restored to Honor, Georgia's B-29 "Sweet Eloise"* and learning new and interesting facts and information. You have my best wishes for its success.

Sincerely,

Roy E. Barnes

REB:pam